Activities Manual

Mathematics for Elementary School Teachers

Ricardo D. Fierro
California State University, San Marcos

Prepared by

Ricardo D. Fierro
California State University, San Marcos

Randa Kress
Idaho State University

BROOKS/COLE
CENGAGE Learning

Australia • Brazil • Japan • Korea • Mexico • Singapore • Spain • United Kingdom • United States

For product information and technology assistance, contact us at
Cengage Learning Customer & Sales Support, 1-800-354-9706

For permission to use material from this text or product, submit all requests online at www.cengage.com/permissions
Further permissions questions can be emailed to permissionrequest@cengage.com

ISBN-13: 978-1-133-36371-2

ISBN-10: 1-133-36371-7

Brooks/Cole
20 Davis Drive
Belmont, CA 94002-3098
USA

Cengage Learning is a leading provider of customized learning solutions with office locations around the globe, including Singapore, the United Kingdom, Australia, Mexico, Brazil, and Japan. Locate your local office at www.cengage.com/global

Cengage Learning products are represented in Canada by Nelson Education, Ltd.

To learn more about Brooks/Cole, visit
www.cengage.com/brookscole

Purchase any of our products at your local college store or at our preferred online store www.cengagebrain.com

Printed in the United States of America
2 3 4 5 6   26 25 24 23 22

Table of Contents

Appendix

About This Activity Manual. This activity manual is designed to accompany the textbook *Mathematics for Elementary School Teachers* by Ricardo D. Fierro. Guiding forces in the development of the manual are: students learn what they have an opportunity to learn; activities should be interesting and worthwhile; activities should correlate with topics covered in the textbook; and time spent exploring, discovering, and understanding the mathematics in the activities should support instruction and assessment.

There is no prescription for how to incorporate the activities in a course as instructors have different teaching styles and approaches, and students have different learning styles. An instructor should use a combination of teacher-directed instruction and student-centered instruction. According to the Mathematics Advisory Panel Report (2008), high-quality research does not support the exclusive use of either of these approaches. Research shows that explicit instruction benefits students. In addition, research also shows that a teacher's knowledge and beliefs "have a strong impact on instructional practice" (Lloyd and Wilson, 1998), so that the approach taken in the college classroom will have a lingering effect upon prospective elementary school teachers, as well as their many future generations of K–6 students. We have assigned some of the activities for use in the classroom, while some of them have been assigned for students to read and complete at home. We believe that these activities can be assigned to accommodate various teaching styles. Students should always have a calculator available for the activities.

The activities are labeled intuitively. For example, Activity 11.2.1 is the first activity that relates to Section 11.2 of the textbook, and Activity 11.2.2 is the second activity that relates to Section 11.2 of the textbook.

Nancy Ressler, a mathematics professor at Oakton Community College in Illinois, deserves special acknowledgment for reviewing the drafts, checking for accuracy, and providing useful comments.

Many thanks to Shaun Williams, Assistant Editor for developmental mathematics at Cengage Learning, for working with us to produce this manual.

Comments and suggestions are welcome, please email them to the authors of the activity manual at fierro@csusm.edu. It would be a pleasure to hear from you.

Best wishes,

Ricardo D. Fierro
Randa Kress

ACTIVITY 1.1.1 Patterns

Material: sharp pencil   Name _____

Patterns and inductive reasoning are stepping stones that students can use for improving their understanding of mathematical principles, problem solving, and algebraic thinking. The following problems give you a chance to be a detective and look for patterns.

1. Fill in the blanks: coincidence, conjecture, guessing, inductive reasoning, opinion, pattern, sequence, or term.

   a. A(n) _____ is similarity or regularity in observations that allows you to predict the behavior in the observations or what comes next.

   b. A(n) _____ is a general statement that seems to be true.

   c. _____ is the process of using patterns to make a conjecture.

   d. A(n) _____ is an ordered arrangement of objects, such as numbers, letters, equations, or shapes.

   e. Each object in a sequence is called a _____.

2. $0 + 1$, $1 + 2$, and $2 + 3$ are examples of sums of two consecutive numbers. In this problem, we will learn a property of the sum of two consecutive numbers.

   a. Find the next three equations that continue the pattern:
   $$0 + 1 = 1$$
   $$1 + 2 = 3$$
   $$2 + 3 = 5$$
   $$3 + 4 = 7$$

   b. Make a conjecture about a property of the sum of two consecutive numbers.

   c. What type of reasoning did you use?

3. 1, 3, 5, and 7 are examples of odd numbers. In this problem, we will learn a property of odd numbers.

   a. Write the next three equations that continue the pattern:
   $$1 = 0 + 1$$
   $$3 = 1 + 2$$
   $$5 = 2 + 3$$
   $$7 = 3 + 4$$

   b. Make a conjecture about a property of odd numbers.

   c. What type of reasoning did you use?

4. Consider the *repeating sequence*
   Q, 7, X, Y, Q, 7, X, Y, …

   a. What patterns do you seen in this sequence?

   b. What is the 12th term in this sequence?

   c. What is the 20th term in this sequence?

   d. What is the 441st term in this sequence?

5. The product of 21 and 5291 equals 111,111.

   a. Find two other whole numbers that have the product 111,111.

   b. Find two whole numbers that have the product 222,222.

   c. Find two whole numbers that have the product 555,555.

6. Use a calculator to determine 35^2, 335^2, 3335^2, and 33335^2. Use a pattern to predict 3333335^2.

7. 4, 10, 16, 22, 28, … is an *arithmetic sequence* with *initial term* 4 and *common difference* 6. List the first five terms of an arithmetic sequence with initial term 3 and common difference 4.

8. 7, 10, 13, 16, 19, … is an arithmetic sequence.

   a. What is the initial term in this sequence?

   b. What is the common difference in this sequence?

   c. What is the 4th term in this sequence?

d. We can *describe the sequence with words* by writing a sentence (fill in the blank): "We begin with _____, and obtain the next term by adding _____ to the previous term."

e. We can *extend the sequence* by listing the next few terms: _____ , _____ , _____ .

f. We can *represent the sequence with a table* by making a table with a few terms. Complete the table.

n, position	1	2			5
y, term	7			16	

g. We can *generalize the sequence* by writing an equation involving the term y and the position n of the term ($n = 1$ corresponds to the 1st term, $n = 2$ corresponds to the 2nd term, and so on). The equation for this arithmetic sequence is $y = 7 + 3(n-1)$ (or $y = 3n + 4$ in simplest form), where y is the term and n is the position of the term. Use this formula to find the 115th term of the sequence.

9. The equation for an arithmetic sequence in simplified form is $y = 4n + 5$.

   a. What is the initial term for this sequence?

   b. What is the common difference for this sequence?

   c. What is the 100th term in this sequence?

ACTIVITY 1.1.2 Limitation of Inductive Reasoning

Material: sharp pencil   Name _____

Elementary students, scientists, mathematicians, and teachers often use inductive reasoning to discover patterns that help them recognize relationships, acquire understanding, and experience some exciting "aha" moments while learning. Are conclusions that are based on a few examples always true?

Below are several circles with n points on circles for n = 2, 3, 4, 5 and chords (line segments whose endpoints are on the circle) connecting the points. The points on the circle are chosen such that no three chords intersect at a single point inside the circle.

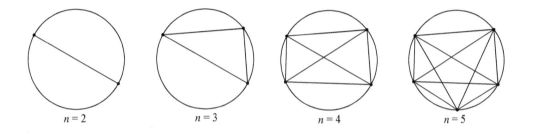

$n = 2$ $n = 3$ $n = 4$ $n = 5$

a. Count the number of non-overlapping regions created by the chords. Complete the table.

n, number of points	2	3	4	5
number of non-overlapping regions	2	4		

b. Look for a pattern in the table, and use the pattern to predict the number of non-overlapping regions for n = 6. Your prediction: _____

c. Now count the number of non-overlapping regions created by the chords constructed from the 6 points (n = 6) on the circle. Actual result: _____

$n = 6$

d. Did your prediction agree with the actual result?

e. What did you learn about inductive reasoning?

ACTIVITY 1.2.1   The Phases of Problem Solving and Four 4's

Material: sharp pencil   Name _____

In this activity you will be exploring Pólya's Problem Solving Process. The four phases of this process allow a person to explore the knowledge that they possess and use it to gain a solution to a problem. The problem presented below needs to be solved; as you solve it, think about what it is that is allowing you to solve it.

Below you will find some equations dealing with four 4's:

$$\frac{4+4+4}{4}=3, \quad \sqrt{4}+\frac{4\cdot4}{4}=6, \quad \text{and} \quad \frac{4}{4}+4+4=9.$$

This shows that the numbers 3, 6, and 9 can be formed with four 4's, square roots, and various operations.
Find a way to form each number between 1 and 25 (see next page) using exactly four 4's and one or more of the
following operations: addition, subtraction, multiplication, division, parenthesis, exponents, decimal points, square
roots, factorials ($4! = 4\cdot3\cdot2\cdot1 = 24$), or any other legal mathematical operation that comes to mind.

Phase 1 of Pólya's problem solving process is "Understand the Problem." At this point, you should reread the problem and look up any words you do not understand. Do you understand what the problem is asking? Is there something that you could do to better understand the problem? Do you understand what you have to do to tackle this problem? Do you understand what form your answers have to take? Can you restate the problem in your own words? When you are confident you understand the problem, you are ready to move to the next phase.

Phase 2 of Pólya's problem solving process is "Devise a Plan." If you are unsure of what strategies to consider, read Section 1.2 of your text. Here you will find a list of possible problem solving strategies that could be used to solve this or any other problem you encounter. What seems to be the best approach for this problem? Does anything from the list seem to fit the parameters stated? Can you think of something to do that is not on the list? Is this problem like any other you have ever solved before? Discuss within your group some possible approaches to the problem.

Phase 3 of Pólya's problem solving process is "Carry Out the Plan." Sometimes, you may discover that your original plan may not work, so you may have to go back and devise another plan. You may even find that you really do not fully understand the problem. If this is true go back and discuss what you are doing right and wrong as a group to find an alternative approach to the problem. Here are some questions you may want to think about: Are you finding a pattern in your solutions? Does there seem to be a trick involved or does the Guess and Check strategy lead you to solutions? Would it help to approach the problem from a different perspective? Are you finding more than one solution for the same number? Is there an operation that you have not thought about or used from the list given within the statement of the problem?

Work with students in a group and list possible solutions below. Share your solutions within the group.

1 _____
2 _____
3 _____
4 _____
5 _____
6 _____
7 _____
8 _____
9 _____
10 _____
11 _____
12 _____
13 _____
14 _____
15 _____
16 _____
17 _____
18 _____
19 _____
20 _____
21 _____
22 _____
23 _____
24 _____
25 _____

Phase 4 of Pólya's problem solving process is "Look Back." Just because a problem is solved does not mean the work is finished. You need to make sure you have answered the question correctly. You may need to check the arithmetic or reasonableness of your answer. You should make sure that your answer fits the information given. This includes rereading the problem again to make sure that what you call a solution really is the solution. Is there another way to represent or model your solution? How could you relate this problem to something you have seen before? Can you pose and solve similar problems? Could you use four 2's or three 3's? This is the most helpful part of look back, to give you a repertoire that could help you solve future problems.

ACTIVITY 1.2.2 Applying Problem Solving Strategies

Material: sharp pencil   Name _____

Use any of the techniques discussed in Section 1.2 of the textbook to solve the following problems. They are designed to make you think. You already have all the reasoning skills needed to solve them. Enjoy!

1. I am thinking of a three digit number. Use the following clues to determine the possible three-digit number.

    The number is under 500.

    The number is odd.

    The sum of the first two digits equals the last digit.

    The number is larger than 200

    Is there only one possible solution to this number riddle?

2. A teacher uses the equation $24 \cdot 4 + 18 \cdot 2 = 132$ to pose a classic barnyard problem: *A farmer wants to know how many horses and ducks are on the farm. He counted 132 legs and 42 animals. How many of each animal is there?*

    a. How does the teacher extract the solution from the equation?

    b. Write your own equation to help you pose and solve a similar barnyard problem.

3. The *digit sum* of a number is the sum of its digits. For example, $67^2 = 4489$ and $4 + 4 + 8 + 9 = 25$, so the digit sum of 67^2 is 25. Find the digit sum of

    a. 53^2.

    b. 533^2.

    c. 533333^2.

4. Mrs. Dugger has 15 coins. She wants to organize them into three piles such that each pile has an odd number of coins. How many ways can this be done?

5. A 3-by-3 *magic square* is table with 3 rows and 3 columns containing the numbers 1, 2, 3, …, 9 such that the sum of every row, column, or diagonal is the same. The common sum is called the "magic number." The magic number of the following magic square is 15.

2	7	6
9	5	1
4	3	8

    a. Suppose you move the first row of the original magic square to the last row as shown. Complete the magic square. What is the magic number?

2	7	6

    b. Suppose you move the corners of the original magic square as shown. Complete the magic square. What is the magic number?

8		6
4		2

    c. Explain why the magic number of any 3-by-3 magic square must be 15.

6. Julie made some cupcakes when she got home from school. Then her brother Ben took half of them to work for his friends and him to eat. When Julie's dad came home from work he ate three before he could stop himself. Then the dog came in the kitchen and knocked four off the counter and then he ate them. When Julie came into the kitchen to frost the cupcakes she found only two left!
   a. How many cupcakes did Julie make?
   b. Which problem solving strategy did you use?

7. Mrs. Smith teaches eighth grade. When she divided her students into groups of 4, she had 3 students leftover. When she divided them into groups of 5, she had 4 students left over. How many students could be in the class?

8. A deli is offering a special for its sandwiches. A customer will receive 2 free sandwiches for every 9 sandwiches purchased. Susie ordered 634 sandwiches for an office party at a large company. How many sandwiches did Susie have to pay for?

9. In the following diagram, the sum of the numbers in the circles equals the number in the square connected to the circles.

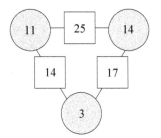

Complete the diagram. What strategy did you use?

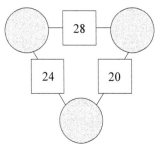

10. Emma is playing a dart board game. The dart board is shown in the diagram.

75	30	50
20	10	5
100	40	60

In this game, she throws darts, one at a time, until two darts have landed in the same square. The number of points is determined by adding up all of the points from the darts. For example, the following throws result in a score of 95.

75	30	50
20	10	5
100	40	60

   a. What are the lowest and highest possible scores?
   b. Suppose Emma throws darts, one at a time, until 3 darts have landed in the same square. What are the lowest and highest possible scores?

11. d is the number of dots in the nth figure. Write an equation that expresses d in terms of n.

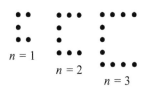

ACTIVITY 1.3.1   Relating Diagrams and Algebra

Material: sharp pencil   Name _____

Algebra is the use of symbols or letters to represent quantities and relationships, and manipulating the symbols and letters. **Algebraic reasoning** is the ability to recognize and solve problems involving unknown quantities. It is a way of thinking. The foundation in algebra that elementary teachers provide is critical to their students' success in later mathematic courses.

1. "Hannah has five fewer than three times as many coins as Cody. Altogether they have 143 coins. How many coins do they each have?" We can use a diagram to represent the relationships.

   a. What does one box represent?
   b. Why is there a vertical arrow with the label 143?
   c. How many coins does the collection of 4 boxes represent?
   d. How many coins does one box represent?
   e. How many coins do they each have?

2. "Hannah has 121 coins. She has seven more than three times as many coins as Cody. How many coins does Cody have?" We can use a diagram to represent the relationships.

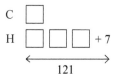

   a. What does one box represent?
   b. Why is there a horizontal arrow with the label 121?
   c. How many coins does the collection of 3 boxes represent?
   d. How many coins does one box represent?
   e. How many coins does Cody have?

Problems 3-5 involve relationships between two quantities. Practice using the method of drawing diagrams explored above and in Section 1.3 of your textbook to represent the relationship. Think about which person is associated with a single "box." Compare your answers to members in your group. Do not use a variable.

3. Robert has twice as many years of experience as Gary.

4. Kitty has three more than twice as many beads as her sister Brittany.

5. Shaneka has two fewer than five times as dolls than Isabella.

Draw a diagram to represent and solve Problems 6-8. When completing your diagram, be able to explain your reasoning at each step. Do not use the Guess and Check strategy or Solve an Equation strategy to solve these problems. Compare your answers to members in your group.

6. Annie has three more apples than Whitney. Together they have 17 apples. How many apples does Whitney have?

7. Alec has 83 songs on his iPod. He has two more than three times as many songs as Collin has on his MP3 player. How many songs does Collin have on his MP3 player?

8. Clint has 11 more than twice as many baseball cards as Lucas. If they have 89 baseball cards altogether, how many baseball cards do they each have?

Use variables to represent relationships in Problems 9-11. Compare your answers to members in your group.

9. Maria has six more than three times as many books as Kelly. Let k represent the number of books Kelly has. Write an expression in terms of k that represents how many books Maria has.

10. Shaun has three fewer than four times as many coins as Marc. Let m represent the number of coins Marc has. Write an expression in terms of m that represents how many coins Shaun has.

11. Krista has two fewer than three times as many friends on Facebook than Carlo. Altogether, they have 94 friends. Write an equation in terms of one variable that can be used to solve for the variable.

Solve Problems 12-14 using the Solve an Equation strategy. Use a single variable. Be able to explain why you chose a specific variable and what it represents. Compare your answers to members in your group.

12. Annie has three more apples than Whitney. Together they have 17 apples. How many apples does Whitney have?

13. Alec has 83 songs on his IPOD touch. He has two more than three times as many songs as Collin has on his MP3 player. How many songs does Collin have on his MP3 player?

14. Clint has 11 more than twice as many baseball cards as Lucas. If they have 89 baseball cards altogether, how many baseball cards do they each have?

15. Write a word problem for each diagram.

    a.   **b.**

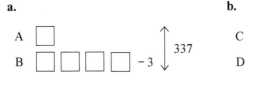

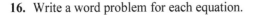

16. Write a word problem for each equation.

    a. $(3n + 10) + n = 366$

    b. $4k - 3 = 225$

ACTIVITY 1.3.2 Is It Magic or Algebra?

Material: sharp pencil   Name _____

In this activity, we will focus on the use of algebra to represent relationships and solve problems.

1. A student thinks of a number, and then follows the directions (see below) given by the teacher. When the student gives the final answer, the teacher tells the student the original number. To students, it may seem like magic. But to the teacher, it is algebra.

    a. Work in a group. One group member should read the following directions for each group member to follow. Use a calculator and press the equal sign after each step.

    Step 1. Pick a counting number; write it on a piece of paper.

    Step 2. Add 7 to the number.

    Step 3. Multiply your answer by 6.

    Step 4. Subtract 4 from your answer.

    Step 5. Divide your answer by 2.

    Step 6. Subtract 1 from your answer.

    Step 7. Divide your answer by 3.

    Step 8. Now tell me your answer.

    b. Compare your original number with your answer. How are they related?

    Did the other members of your group experience the same relationship?

    c. Represent the directions with algebraic expressions. *Do not simplify the expressions.*

Directions	Algebraic expression
Step 1. Pick a counting number, write it on a piece of paper.	Step 1. n
Step 2. Add 7 to the number.	Step 2.
Step 3. Multiply your answer by 6.	Step 3.
Step 4. Subtract 4 from your answer.	Step 4.
Step 5. Divide your answer by 2.	Step 5.
Step 6. Subtract 1 from your answer.	Step 6.
Step 7. Divide your answer by 3	Step 7.

    d. Simplify the final algebraic expression in the table in Step 7.

    e. Did the simplified expression help explain your observation from part (b)?

    f. Suppose a student's answer was 67. What was the student's original number?

2.   Follow these directions. Use a calculator and press the equal sign after each operation.

Step 1.   Think of a counting number.

Step 2.   Double the number.

Step 3.   Add 30 to the result.

Step 4.   Divide the result by 2.

Step 5.   Subtract 6 from the result.

a.   Describe how the final result the calculator displays is related to your original number.

b.   Share your original number and the result displayed on the calculator with another member of your group.

c.   Represent the directions using algebra. Write the algebraic expressions that are described within each step. Use *n* for the original number. Simplify the expression at each step.

Directions	Algebraic expression
Step 1. Think of a counting number.	Step 1. *n*
Step 2. Double the number.	Step 2.
Step 3. Add 30 to the result.	Step 3.
Step 4. Divide the result by 2.	Step 4.
Step 5. Subtract 6 from the result.	Step 5.

3.   Follow these directions. Use a calculator and press the equal sign after each operation.

Step 1.   Think of a counting number.

Step 2.   Add 7 to the number.

Step 3.   Double the result.

Step 4.   Subtract 10 from the result.

Step 5.   Divide the result by 2.

Step 6.   Announce the final result.

Use algebra to determine how a magician would determine the final result from the number that was announced.

4.   The table shows a calendar for September 2013 (a calendar for any month of any year would work for this problem, but one is supplied for your for convenience). A magician asks you to pick any 2-by-2 array of days (such as 1, 2, 8, and 9) and asks you to announce the sum of the four days. The magician thinks for a moment and then tells you the four days that you picked. How is this possible?

September 2013						
Su	Mo	Tu	We	Th	Fr	Sa
1	2	3	4	5	6	7
8	9	10	11	12	13	14
15	16	17	18	19	20	21
22	23	24	25	26	27	28
29	30					

5.   Design a magic trick that could be shared with the class.

Make sure to include the algebraic expressions that show why the trick works.

ACTIVITY 2.1.1   Sets and Operations

Material: sharp pencil   Name _____

A **set** is a well defined collection of objects. We typically name sets using capital letters. $\{1, a, k\}$ is read "the set containing 1, a, and k." $A = \{1, a, k\}$ is read "A is the set containing 1, a, and k." Sets give mathematicians a way to group objects or visualize relationships.

1. The *universal set* is the set of all objects under consideration. Find all solutions to the equation $x^2 = 25$.
    Assume the universal set U is

    a. $U = \{0, 1, 2, 3, \ldots\}$

    b. $U = \{\ldots, -3, -2, -1, 0, 1, 2, 3, \ldots\}$

    c. $U = \{2, 4, 6, 8, 9, \ldots\}$

2. Evaluate $11 \div 4$. Assume the universal set is the set of

    a. whole numbers.

    b. fractions.

    c. real numbers.

3. Refer to the Venn diagram shown. Write each set using set notation (such as $\{3, 6, 7\}$).

    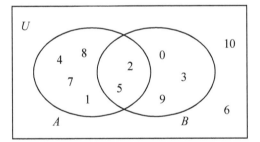

    a. Set A.

    b. Set B.

    c. The universal set U.

    d. All objects in A or B, or both.

    e. All objects in both A and B.

    f. All objects in A, but not B.

    g. All objects in B, but not A.

    h. All objects that are not in A.

    i. All objects that are not in B.

4. Shade the portion of the Venn diagram of A, B, and the universal set U that represents the set.

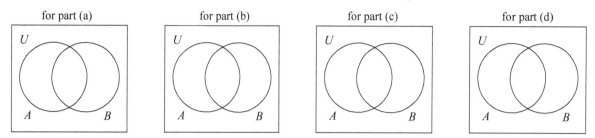

for part (a)   for part (b)   for part (c)   for part (d)

a. $A \cup B$, read "A union B," which is the set of all objects that belong to A or B, or both.

b. $A \cap B$, read "A intersect B," which is the set of all objects that belong to A and B.

c. $A - B$, read "the difference of A and B," which is the set of all objects that belong to A, but not B.

d. $\sim A$, read "the complement of A," which is the set of all objects that do not belong to A.

5. The universe is $U = \{0, 1, 2, 3, 4, 5, 6, 7, 8\}$. Determine $A \cup B$, $A \cap B$, $A - B$, $B - A$, $\sim A$, and $\sim B$.

a. $A = \{1, 3, 5, 7\}$ and $B = \{3, 5, 6\}$

b. $A = \{3, 6, 1\}$ and $B = \{1, 7\}$

6. Let A be the set of students who completed their assignments and let B be the set of students who passed the class. Interpret each set: $A \cup B$, $A \cap B$, $A - B$, $B - A$

7. Two sets A and B are *equivalent*, written $A \sim B$, if and only if there is a one-to-one correspondence between their elements. For example, $A = \{1, k, 2, \&\}$ and $B = \{0, 7, 4, +\}$ are equivalent because we can match their elements with the one-to-one correspondence $1 \leftrightarrow 0$, $k \leftrightarrow 7$, $2 \leftrightarrow 4$, and $\& \leftrightarrow +$. The sets $A = \{1, 2, 3, 4, 5, \dots\}$ and $B = \{3, 6, 9, 12, 15, \dots\}$ are equivalent. If you pick the element n of A, what could be the corresponding element of B?

8. Draw a Venn diagram of A, B, and the universal set U. Let $n(C)$ denote the number of objects in the set C. Determine $n(A)$, $n(B)$, $n(A \cup B)$, $n(A \cap B)$. Is $n(A \cup B) = n(A) + n(B)$? Why or why not? If not, what formula would work?

a. $A = \{a, b, c, d\}$, $B = \{b, e\}$, and $U = \{a, b, c, d, e, f, g\}$.

b. $A = \{a, b, c\}$, $B = \{a, b, e, f, g\}$, and $U = \{a, b, c, d, e, f, g\}$.

9. How many objects does the set $\{4, 2, \{1, 3, 5, 7, \dots\}\}$ have?

10. The symbol $A \subseteq B$, which is read "A is a subset of B," means every object in A also belongs to B. For example, if $A = \{a, b, c\}$ and $B = \{a, b, c, d, e\}$, then $A \subseteq B$. However, $B \nsubseteq A$ because e belongs to B but e does not belong to A. For the given sets A and B, is $A \subseteq B$?

a. $A = \{1, 2, 3, 6\}$ and $B = \{1, 4, 8, 5, 6, 3, 2\}$

b. $A = \{4, 8, 9\}$ and $B = \{1, 4, 8\}$

c. $A = \{\ \}$ and $B = \{1, 5, 2\}$

ACTIVITY 2.1.2 Representing and Solving Counting Problems with Venn Diagrams

Material: sharp pencil   Name _____

In this activity, you will learn how to use Venn diagrams to organize information and solve logical reasoning problems involving counting.

1. Mrs. Dugger surveyed her students. She asked them if they like M&Ms or Skittles.
    The results are summarized in the Venn diagram shown.

    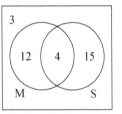

    a. How many students like M&Ms?

    b. How many students like Skittles?

    c. How many students like both M&Ms and Skittles?

    d. How many students like only M&Ms?

    e. How many students like only Skittles?

    f. How many students like just one candy?

    g. How many students like neither candy?

    h. How many students were surveyed?

2. Mr. Brown surveyed his class of 25 students. In the survey, 12 students said they have a dog,
    7 students said they have a cat, and 2 students said they have both a cat and a dog.

    a. Represent the information with a Venn diagram.

    b. How many students have only a dog or only a cat?

    c. How many students do not have a dog or cat?

3. Students who had seen either an eclipse or meteor were surveyed.
    Of these students, 16 students saw an eclipse and 25 students saw a meteor.

    a. What is the minimum number of students surveyed?

    b. What is the maximum number of students surveyed?

4. Mrs. Smith surveyed her students. She asked them if they had been to
    Disneyland, Legoland, or Sea World. Let *D* be the set of students have
    been to Disneyland, let *L* be the set of students who have been to
    Legoland, and let *S* be the set of students who have been to Sea World.
    Match the shaded region with the interpretations of the shaded regions.

    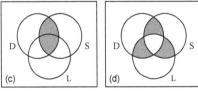

    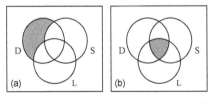

    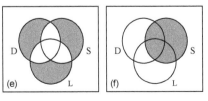

    (i)   students who have been to only one of the three theme parks

    (ii) students who have been to exactly two theme parks

    (iii) students who have to all three theme parks

    (iv) students who have been to Disneyland and Sea World

    (v)   students who have been to Sea World

    (vi) students who have only been to Disneyland

5. Mr. Roberts surveyed students who like hop scotch, jump rope, or frisbee.

    The results of the survey are shown in the Venn diagram.

    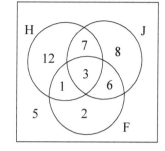

    a. How many students liked all three activities?

    b. How many students liked exactly one activity?

    c. How many students liked exactly two activities?

    d. How many students liked hop scotch or jump rope?

    e. How many students only liked hop scotch and jump rope?

    f. How many students were surveyed?

    g. How many students liked jump rope but not frisbee?

6. Mrs. Rocha surveyed 35 students who liked tennis, running, or skateboarding.

    - 5 students said they liked tennis, running, and skateboarding.

    - 8 students said they liked running and skateboarding.

    - 11 students said they liked tennis and running.

    - 10 students said they only liked skateboarding.

    - 7 students said they only liked tennis or only liked running.

    How many students liked exactly two of these activities?

7. 37 customers at a local restaurant who ordered drinks, dinner, or dessert volunteered to participate in a survey.

    The data showed:

    - 5 customers ordered drinks, dinner, and dessert.

    - 9 customers ordered drinks and dinner.

    - 6 customers ordered only dessert.

    - 7 customers ordered dinner and dessert.

    - 21 customers ordered dessert.

    - 12 customers ordered only drinks or only dinner.

    a. How many customers ordered exactly two items?

    b. How many customers ordered exactly one item?

8. Write a counting problem that uses the Venn diagram shown. Pose several questions.

    Ask someone in your group to answer the questions. Then verify their answers.

    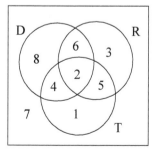

ACTIVITY 2.2.1 Place Value

Material: sharp pencil   Name _____

A solid understanding of place value concepts is an essential goal that students should achieve in the primary grades K-2. In this activity you will learn and apply some basic place value concepts, which form the foundation for success number sense and standard addition, subtraction, multiplication, and division methods.

1. The diagrams below show two ways to represent 32: 3 tens, 2 ones and 2 tens, 12 ones.

    a. Sketch another representation of 32.

    b. What is the minimum number of pieces needed to represent 32?

    c. What is the maximum number of pieces needed to represent 32?

    d. How many ways can you represent 32 using base ten blocks?

2. The diagrams below show two ways to represent 46: 4 tens, 6 ones and 3 tens, 16 ones.

    a. What is the minimum number of pieces needed to represent 46?

    b. What is the maximum number of pieces needed to represent 46?

    c. How many ways can you represent 46 using base ten blocks?

3. Fill in the blanks. Use the most efficient representation possible (the minimum number of base ten pieces possible) so that equality holds. Check your answers with another student.

    a. 345 = _____ tens _____ ones

    b. 2468 = _____ hundreds _____ tens _____ ones

    c. 4731 = _____ thousands _____ ones

    d. 18,054 = _____ hundreds _____ tens _____ ones

4. The following sketch is a place value model of 345. Sketch a model of 427.

5. The *place value* of a digit indicates the particular base ten block that is repeated. The *digit* in a numeral indicates how often the base ten block is repeated. The *value* of a digit indicates how much the digit is worth by combining the digit and the place value. For example, in the numeral 456, the place value of the 4 is hundreds and the value of the 4 is 400. Find the place value and value of each digit.

    a. 353

    b. 72,029

6. The short word form of 34,567,521 is 34 million, 567 thousand, 521. Write the short word form of each number.

   a. 45,083

   b. 6,507,362

7. The *expanded form* of 3035 is $3000 + 0 + 30 + 5$. The expanded form gives us a way to represent 3035, for example, in terms of the values of the digits. Write the expanded form of each number.

   a. 5678

   b. 6604

8. The expanded form of 3035 in terms of multiplication is $3 \cdot 1000 + 0 \cdot 100 + 3 \cdot 10 + 5 \cdot 1$. This form gives us a way to represent 3035, for example, in terms of the place values of the digits. Write the expanded form of each number in terms of multiplication.

   a. 5678

   b. 6604

9. Multiply the following numbers: 39, your age, and 259. Explain why you obtained those results.

10. This magic trick relies on place value.

    a. Use a calculator and press the equal sign after each operation.

    Step 1. Enter the month of your birthday (Jan = 1, Feb = 2, ..., Dec = 12).
    Step 2. Double that number.
    Step 3. Add 40.
    Step 4. Multiply by 10.
    Step 5. Subtract 7.
    Step 6. Multiply by 5.
    Step 7. Add the number that is the day of your birthday.
    Step 8. Subtract 1965.

    b. Describe what the calculator is displaying. Share your results with another student.

    c. Represent the directions using algebra. Write the algebraic expressions that are described within each step.

    Use *m* for the month and *d* for the day. Simplify the expression at each step.

directions	algebraic expression
Step 1. Enter the month of your birthday.	Step 1. *m*
Step 2. Double that number.	Step 2.
Step 3. Add 40.	Step 3.
Step 4. Multiply by 10.	Step 4.
Step 5. Subtract 7.	Step 5.
Step 6. Multiply by 5.	Step 6.
Step 7. Add the number that is the day of your birthday	Step 7.
Step 8. Subtract 1965.	Step 8.

    d. Why is place value critical for this problem?

ACTIVITY 2.3.1 Representations of Addition and Subtraction

Material: sharp pencil   Name _____

Additive reasoning is the ability to recognize and solve problems involving addition and subtraction. In this activity we focus on helping you recognize contextualized situations that involve addition and subtraction. We also show you various ways to solve one-step equations such as $n + 36 = 82$.

- "Maria has 3 coins. Jaime has 5 coins. How many coins do they have altogether?" This problem follows the *set model* of addition. We combine two discrete sets of objects (such as coins or marbles), making sure the sets have no objects in common, and the sum 3 + 5 represents the total number of objects in the sets combined.
- "Lesley has 8 coins. She gave Alex 3 coins. How many coins does she have left?" This problem follows the *take away model* of subtraction. We have a collection of objects and remove a subset of the objects. In this problem, the difference 8 – 3 represents the number of coins Lesley has left.
- "Coco has 10 dollars. She needs $30 to buy a computer game. How much more money does she need?" This problem follows the *missing addend model* of subtraction because the underlying equation is $5 + n = 30$ and the difference 20 – 5 reveals how much more money Coco needs.
- Krista has 15 books and Carlo has 12 books. How many fewer books does Carlo have?" This problem follows the *comparison model* of subtraction because we are comparing two given quantities and the difference 15 – 12 tells how many fewer books Carlo has. We could have asked "How many more books does Krista have?"

1. *Classify* the word problem using one the following models: take away model, missing addend model, and comparison model. Then explain why you chose the model.
    a. Pamela has 75 beads. Kari has 68 beads. How many more beads does Pamela have than Kari?
    b. The grocer put 15 oranges in the box. The box can hold 35 oranges. How many more oranges can the grocer put in the box?
    c. There are 24 books on a shelf. Jo Anne gave 12 books to teachers. How many books remain on the shelf?

Some students look for key words to identify the operation in a word problem.
2. Consider the following word problem. "Korbi has 35 coins. She has 3 more coins than Holly. How many coins does Holly have?"
    a. A student thinks the answer is 38 because $35 + 3 = 38$. Why do you think the student added?
    b. Draw a diagram that could help illustrate the relationships in the word problem and help the student solve the problem correctly.

3. Consider the following problem. "Benny has 46 coins. Benny has 5 fewer coins than Dallas. How many coins does Dallas have?"
    a. A student thinks the answer is 41 because $46 - 5 = 41$. Why do you think the student subtracted?
    b. Draw a diagram that could help illustrate the relationships in the word problem and help the student solve the problem correctly.

4. Here's the definition of subtraction: Let a and b represent whole numbers such that $a \geq b$.

    Then $a - b$, read a minus b, is the whole number n (that is, $a - b = n$) if and only if $a = b + n$.

    Note: $a - b = n$ if and only if $a = n + b$, too, by the commutative property of addition. We can solve various

    equations using the definition of subtraction:

$n - 17 = 31$	$18 + n = 53$	$12 - n = 5$
$n = 31 + 17$ *by the defn. of subtraction*	$n = 53 - 18$ *by the defn. of subtraction*	$12 = 5 + n$ *by the defn. of subtraction*
$n = 48$	$n = 35$	$12 - 5 = n$ *by the defn. of subtraction*

    Solve each equation using the definition of subtraction.

    a. $n - 26 = 64$

    b. $28 + n = 72$

    c. $13 - n = 8$

5. Use a calculator. Enter the $=$ sign after each operation.

    a. Enter 56. Add 34. Subtract 34. What is your answer?

    b. Enter 75. Add 49. Subtract 49. What is your answer?

    c. Use you examples to complete the following equation: $a + b - b = $ ___ .

    d. Enter 83. Subtract 24. Add 24. What is your answer?

    e. Enter 52. Subtract 40. Add 40. What is your answer?

    f. Use you examples to complete the following equation: $a - b + b = $ ___ .

    g. Enter 30. Add 18. Subtract 30. What is your answer?

    h. Enter 75. Add 21. Subtract 75. What is your answer?

    i. Use you examples to complete the following equation: $a + b - a = $ ___ .

    j. Read Example 2.35 in your textbook. What does it prove?

    The equations $a + b - b = a$, $a - b + b = a$, and $a + b - a = b$ represent the idea that addition and subtraction are

    inverse operations. Here's how we can solve equations using the fact that addition and subtraction are inverse

    operations.

$n - 53$	$=$	82	*original equation*
$n - 53 + 53$	$=$	$82 + 53$	*addition property of equality*
n	$=$	$82 + 53$	*addition and subtraction are inverse operations*
n	$=$	135	*simplification*

6. Use the fact addition and subtraction are *inverse operations* to solve the equation.

    a. $n + 23 = 47$

    b. $k - 42 = 75$

    c. $w + 30 = 62$

    d. $m - 20 = 55$

    e. $12 - n = 5$

ACTIVITY 2.3.2 Number Sense with Addition and Subtraction

Material: sharp pencil   Name _____

Number sense is a way of thinking about numbers, relationships between numbers, and relationships between and among operations. A student with number sense knows various ways to represent 28, such as $28 = 2$ tens, 8 ones $= 1$ ten, 18 ones, $28 = 20 + 8$, and $28 = 30 - 2$. Before students learn their addition table and become fluent with addition facts such as $8 + 5 = 13$, they should develop at least one strategy for each addition fact. Fluency with basic addition facts such as $8 + 5 = 13$ is needed for more advanced problems such as $468 + 75$. In this activity, we will explore various ways to add and subtract that rely on number sense so that you will be prepared to give your students learning opportunities to explore strategies for learning basic addition and subtraction facts. Along the way, your students will see that these facts are related rather than independent.

1. We can use the counting on strategy to add: 8 + 5. First, say the number 8. Then count five more by saying "nine, ten, eleven, twelve, thirteen." Then $8 + 5 = 13$. Use the counting on strategy to add.

   a. $7 + 5$

   b. $9 + 4$

2. Suppose a student uses the counting on strategy to add. Explain how the commutative property would be useful for $3 + 8$.

3. Students often know their "doubles" such as
   $$1 + 1 = 2$$
   $$2 + 2 = 4.$$
   $$3 + 3 = 6$$
   a. What are the next two equations in the sequence?

   b. A diagram for the double 3 + 3 is shown: ⦂ ⦂ ⦂ . Draw a diagram for the double 5 + 5.

4. Students can build on their knowledge of doubles to add. For example, $8 + 5 = 3 + 5 + 5 = 3 + 10 = 13$. With this strategy, students relate the unknown sum 8 + 5 to the related and known double 5 + 5. Use the strategy of using doubles to add.

   a. $6 + 7$

   b. $9 + 7$

5. We can also find a sum by decomposing (or breaking apart or compensation) an addend to make an easy combination such as making ten. For example, we can break apart the 5 to make 10: $8 + 5 = 8 + 2 + 3 = 10 + 3 = 13$. The making ten strategy builds on the students knowledge of the various ways to form ten: $9 + 1 = 10$, $8 + 2 = 10$, $7 + 3 = 10$, $6 + 4 = 10$, and $5 + 5 = 10$. The strategy is an important bridge to mastering basic addition facts. Use the strategy of making ten to add.

   a. $9 + 8$

   b. $7 + 5$

6. Students quickly learn the pattern that arises in adding 10 to a number. For example, $4+10=14$, $18+10=28$, and $10+37=47$. Discuss how you could exploit this to add 9, such as in the problems $9+7$, $42+9$, and $9+83$.

7. The take away model of subtraction is a concrete way to help students think of subtraction. Let's consider the problem $7-4$. We begin by representing the minuend 7 with seven objects. Then we take away 4 objects. The difference $7-4$ equals the number of objects remaining. Then $7-4=3$.

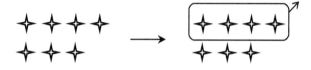

Use the take away model to subtract.

a. $9-4$

b. $8-6$

8. Another way to think of subtraction is the "missing addend" model. Students turn the equation $7-4=$ ___ into the equation $7=4+$ ___ . Students know that $7=4+3$, so $7-4=3$. Apply the missing addend model to subtract:

a. $13-7$

b. $15-8$

9. Students quickly learn the pattern that arises in subtracting 10 from a number. For example, $27-10=17$, $38-10=28$, and $62-10=52$. Discuss how you could use 10 as a bridge in subtraction to develop a thinking strategy to subtract

a. 9, such as in the problems $46-9$, $73-9$, and $36-9$.

b. 8, such as in the problems $35-8$, $42-8$, and $63-8$.

10. Another way to subtract hinges on decomposition. For example, we can think of $14-6$ as $14-4-2$ (to take away 6, first take away 4, then take away 2). Then $14-4=10$ and $10-2=8$. So $14-6=8$. Use decomposition to subtract.

a. $12-7$

b. $15-8$

11. The following diagram suggests a possible way to subtract: $53-28$ by counting forward. A student draws an "empty number line" (no tick marks), and then uses natural counting strategies and counts *forward* to obtain $2+20+3=25$. Then $53-28=25$.

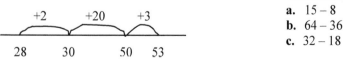

Use the empty number line to subtract.

a. $15-8$

b. $64-36$

c. $32-18$

12. Learning the addition table is a critical goal in the primary grades K-2. In this problem, we will see that patterns are useful in learning addition facts.

+	0	1	2	3	4	5	6	7	8	9
0	0	1	2	3	4	5	6	7	8	9
1	1	2	3	4	5	6	7	8	9	10
2	2	3	4	5	6	7	8	9	10	11
3	3	4	5	6	7	8	9	10	11	12
4	4	5	6	7	8	9	10	11	12	13
5	5	6	7	8	9	10	11	12	13	14
6	6	7	8	9	10	11	12	13	14	15
7	7	8	9	10	11	12	13	14	15	16
8	8	9	10	11	12	13	14	15	16	17
9	9	10	11	12	13	14	15	16	17	18

   a. Shade the squares that involve addition with zero.

       Which property allows students to know $0 + 3 = 3$ and $5 + 0 = 5$?

   b. To find $n + 1$, students know that it is the next counting number.

       Shade the squares that involve adding one more.

   c. To find $n + 2$, students know that it is the next even number or next odd number. For example, $6 + 2 = 8$ because 8 is the next even number after 6. Shade the squares that involve adding two more.

   d. Doubles (such as 6 + 6) follow a pattern and give students a platform for other combinations (such as 6 + 7). Shade the squares that involve doubles.

   e. Students can use doubles to find "near doubles" such as 7 + 6 by thinking $7 + 6 = 1 + 6 + 6 = 1 + 12 = 13$. Shade the squares that involve these types of sums where the addends differ by 1.

   f. Students can add 9 by adding 10 and then subtracting 1. Shade the squares that involve adding 9.

   g. Suppose a student knows $a + b$. Which property allows a student to quickly determine $b + a$? Shade the squares in the upper portion of the table that allows students to use this property.

   h. Students should learn different ways to make 10, such as $8 + 2 = 10$ and $7 + 3 = 10$. Shade the squares that involve making 10.

   i. Count the number of unshaded squares in the table. There should be 8 unshaded squares. If you have fewer or more than 8 unshaded squares, then review parts (a)-(h) and erase or shade squares as needed so that there are 8 unshaded squares. Of the 100 addition facts, it appears that 8 of them require some special attention.

ACTIVITY 2.4.1   Making Sense of the Standard Addition Algorithm

Material: sharp pencil and base ten blocks (p. A1)   Name _____

In this activity you will use base ten blocks to make sense of the standard addition algorithm. Hopefully you will see that the word "carry" that you may currently does not accurately reflect what happens when elementary school students use base ten blocks to add. The word "regroup" would be more appropriate when making the connection between base ten blocks and the standard addition algorithm.

Use base ten blocks to add: $26 + 38$

Step 1. Represent each addend with base ten blocks.

Step 2. Add the ones and regroup as necessary:
6 ones + 8 ones = 14 ones = 1 ten, 4 ones

Step 3. Add the tens. 1 ten + 2 tens + 3 tens = 6 tens. No more regrouping is needed. This leaves 6 tens and 4 ones, so $26 + 38 = 64$.

The **standard addition algorithm** mimicks the actions with the base ten blocks.

Step 1. Use a vertical format. Align the place values.

```
  2 6
+ 3 8
-----
```

Step 2. Add the ones: 6 ones + 8 ones = 14 ones = 1 ten, 4 ones. The 1 above the 2 represents the 1 ten from regrouping the 14 ones as 1 ten, 4 ones.

```
  1
  2 6
+ 3 8
-----
    4
```

Step 3. Add the tens: 1 ten + 2 tens + 3 tens = 6 tens. No more regrouping is needed. So $26 + 38 = 64$.

```
  1
  2 6
+ 3 8
-----
  6 4
```

Note: The 1s above digits in the top addend in the standard addition algorithm for addition indicate *regroupings*. Also, the word "carry" was never used in describing the standard addition algorithm.

---

1.   Use base ten blocks to add 285 and 79. Then answer the following questions.

```
  1 1
  2 8 5
+   7 9
-------
  3 6 4
```

   a. Explain why the 1 appears above the 8.

   b. Explain why the 1 appears above the 2.

2. Use base ten blocks to add 674 and 269. Then answer the following questions.

$$
\begin{array}{r}
\overset{1}{} \overset{1}{} \\
6\ 7\ 4 \\
+\ 2\ 6\ 9 \\
\hline
9\ 4\ 3
\end{array}
$$

  a. Explain why the 1 appears above the 7.

  b. Explain why the 1 appears above the 6.

3. A student adds the two numbers $456 + abc$ using the standard addition algorithm, where a, b, and c are nonzero digits. Determine the largest possible number abc such that

  a. no regroupings are required.

  b. exactly one regrouping is required.

  c. exactly two regroupings are required.

4. Do the following.

  a. A student adds the two numbers $4a$ and $b8$ as shown.
    What are the values of a and b?

  b. A student adds the two numbers uv and wx as shown.
    What is the largest possible value for k?

$$
\begin{array}{r}
\overset{1}{} \\
4\ a \\
+\ b\ 8 \\
\hline
8\ 4
\end{array}
\qquad
\begin{array}{r}
k \\
u\ v \\
+\ w\ x \\
\hline
y\ z
\end{array}
$$

5. Give an example of an addition problem in which it is useful to know both
12 ones = 1 ten, 2 ones and 15 tens = 1 hundred, 5 tens.

6. A student adds the two numbers $25a$ and $3b6$.
What are the values of a, b, and c?

$$
\begin{array}{r}
2\ 5\ a \\
+\ 3\ b\ 6 \\
\hline
c\ 5\ 4
\end{array}
$$

7. Find the missing digits.

$$
\begin{array}{r}
8\ \square\ 3 \\
+\ \square\ 4\ \square \\
\hline
\square\ 3\ 3\ 2
\end{array}
$$

8. A student adds two numbers as shown.

$$
\begin{array}{r}
\overset{1}{} \\
6\ 7 \\
+\ 2\ 5 \\
\hline
9\ 2
\end{array}
$$

The students says: "seven plus five equals twelve, write the two, carry the one, one plus six plus two equals nine, write the nine. The sum is ninety-two."

  a. What word is more appropriate than "carry?"

  b. When the student adds seven and five, they are adding some "things". What are they adding?

  c. When the student adds one, six, and two, they are adding some "things". What are they adding?

ACTIVITY 2.4.2   Making Sense of the Standard Subtraction Algorithm

Material: sharp pencil and base ten blocks (p. A1)   Name _____

In this activity you will use base ten blocks to make sense of the standard subtraction algorithm. Hopefully you will see that the word "borrow" that you may currently use does not accurately reflect what happens when elementary school students use base ten blocks to subtract. The word "regroup" would be more appropriate when making the connection between base ten blocks and the standard subtraction algorithm.

Use base ten blocks to subtract: $64 - 38$.

Step 1. Represent the minuend 64 with base-ten blocks.

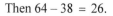

Step 2. There are 4 ones, but we need to take away 8 ones. We regroup one ten as 10 ones, leaving 5 tens. Then 10 ones + 4 ones = 14 ones.

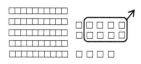

Step 3. Now there are enough ones to take away 8 ones. This leaves 4 ones.

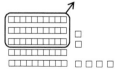

Step 4. There are 5 tens, and we take away 3 tens. This leaves 2 tens.

Step 5. There are 2 tens and 6 ones.

Then $64 - 38 = 26$.

The **standard subtraction algorithm** mimics the actions with the base ten blocks.

Step 1. Use a vertical format. Align the place values.

$$\begin{array}{r} 6\ 4 \\ -\ 3\ 8 \\ \hline \end{array}$$

Step 2. There are 4 ones, but we need to take away 8 ones. We regroup one ten as 10 ones, leaving 5 tens. Then 10 ones + 4 ones = 14 ones.

$$\begin{array}{r} \overset{5}{\cancel{6}}\ \overset{14}{\cancel{4}} \\ -\ 3\ 8 \\ \hline \end{array}$$

Step 3. Now there are enough ones to take away 8 ones. This leaves 4 ones.

$$\begin{array}{r} \overset{5}{\cancel{6}}\ \overset{14}{\cancel{4}} \\ -\ 3\ 8 \\ \hline 6 \end{array}$$

Step 4. There are 5 tens, and we take away 3 tens. This leaves 2 tens.

$$\begin{array}{r} \overset{5}{\cancel{6}}\ \overset{14}{\cancel{4}} \\ -\ 3\ 8 \\ \hline 2\ 6 \end{array}$$

Note: The crossed out digits in the standard subtraction algorithm indicate *regroupings*. Also, we never used the word "borrow" to describe the standard subtraction algorithm.

1. Work in pairs. Use base ten blocks to subtract. Explain the steps

    to your partner. Avoid using the term "borrow."

    a. $75 - 38$

    b. $437 - 59$

    c. $523 - 367$

2. Use base ten blocks to subtract: $374 - 125$. Then answer the following questions.

    $$
    \begin{array}{r}
    3 \; \overset{6}{\cancel{7}} \; \overset{14}{\cancel{4}} \\
    - \; 1 \;\; 2 \;\; 5 \\
    \hline
    2 \;\; 4 \;\; 9
    \end{array}
    $$

    a. Explain why the 6 appears above the 7.

    b. Explain why the 14 appears above the 4.

3. Use base ten blocks to subtract: $653 - 115$. Then answer the following questions.

    $$
    \begin{array}{r}
    6 \; \overset{4}{\cancel{5}} \; \overset{13}{\cancel{3}} \\
    - \; 1 \;\; 1 \;\; 5 \\
    \hline
    5 \;\; 3 \;\; 8
    \end{array}
    $$

    a. Explain why the 4 appears above the 5.

    b. Explain why the 13 appears above the 3.

4. Find the missing digits.

    $$
    \begin{array}{r}
    6 \;\; \square \;\; 7 \\
    - \; 2 \;\; 5 \;\; \square \\
    \hline
    \square \;\; 8 \;\; 9
    \end{array}
    $$

5. Find the missing digits.

    $$
    \begin{array}{r}
    6 \;\; \square \;\; 8 \\
    - \; \square \;\; 5 \;\; \square \\
    \hline
    4 \;\; 7 \;\; 5
    \end{array}
    $$

6. A student subtracts $675 - abc$ using the standard subtraction algorithm, where a, b, and c are nonzero digits.

    Determine the largest possible number abc such that

    a. no regroupings are required.

    b. one regrouping is required.

    c. two regroupings are required.

7. A student subtracts $84a - 3b6$ as shown, where a, b, and c
    are nonzero digits. What are the values of a, b, and c?

    $$
    \begin{array}{r}
    8 \;\; 4 \;\; a \\
    - \; 3 \;\; b \;\; 6 \\
    \hline
    c \;\; 5 \;\; 4
    \end{array}
    $$

8. Give an example of a subtraction problem in which it is useful to know $34 = 2$ tens, 14 ones.

ACTIVITY 3.1.1 Properties of Multiplication and Number Sense

Material: sharp pencil   Name _____

Properties of multiplication give students an early way to determine products such as $6 \cdot 7$ or $7 \cdot 19$ before they learn the standard multiplication algorithm. Properties of whole numbers also provide the foundation for number sense, estimation skills, and algebra.

a. The commutative property of multiplication is useful for early learning of multiplying by 2 and 5, since students are familiar with doubling and counting by fives. For example, $7 \cdot 2 = 2 \cdot 7 = 14$ (double: 7, 14) and

$5 \cdot 3 = 3 \cdot 5 = 15$ (count by fives: 5, 10, 15).

b. The associative property of multiplication is useful for simplifying some calculations. For example,

$25 \cdot 12 = 25 \cdot (4 \cdot 3) = (25 \cdot 4) \cdot 3 = 100 \cdot 3 = 300$.

c. The distributive property of multiplication over addition is useful for finding some products using known multiplication facts. For example, $8 \cdot 12 = 8 \cdot (10 + 2) = 8 \cdot 10 + 8 \cdot 2 = 80 + 16 = 96$ and

$12 \cdot 8 = (10 + 2) \cdot 8 = 10 \cdot 8 + 2 \cdot 8 = 80 + 16 = 96$. This property forms the basis for the standard algorithm for multiplication.

d. The distributive property of multiplication over subtraction is useful for finding some products using known multiplication facts. For example, $15 \cdot 9 = 15 \cdot (10 - 1) = 15 \cdot 10 - 15 \cdot 1 = 150 - 15 = 135$.

e. The properties form the basis for simplifying algebraic expressions. For example, we know $3a + 5a = 8a$. Here's why: $3a + 5a = (3 + 5)a = 8a$, because we applied the distributive property of multiplication over addition.

For problem 1-10, use number sense to multiply. This means you have to look at the factors and then decide how to proceed to multiply the numbers. Share your strategy with another student. Compare your strategies.

1. $15 \cdot 20$

2. $11 \cdot 42$

3. $18 \cdot 17$

4. $35 \cdot 95$

5. $60 \cdot 40 =$

6. $65 \cdot 18$

7. $20 \cdot 45$

8. $567 \cdot 2$

9. $25 \cdot 24$

10. $4 \cdot 27$

11. On a recent exam, several students wrote $(3a)^2 + (2a)^2 = 5a^2$.

How would you use the meaning of exponents to correct their thinking?

ACTIVITY 3.1.2 Learning Basic Multiplication Facts

Material: sharp pencil   Name _____

Learning the multiplication table represents a milestone for elementary school students. Fluency with basic multiplication facts such as $3 \times 5 = 15$ is needed to form more complex products such as 245×38. Hopefully, after this activity you will realize that the challenge of learning the 100 basic multiplication facts in the multiplication table shown below is not as formidable as it may seem.

×	0	1	2	3	4	5	6	7	8	9
0	0	0	0	0	0	0	0	0	0	0
1	0	1	2	3	4	5	6	7	8	9
2	0	2	4	6	8	10	12	14	16	18
3	0	3	6	9	12	15	18	21	24	27
4	0	4	8	12	16	20	24	28	32	36
5	0	5	10	15	20	25	30	35	40	45
6	0	6	12	18	24	30	36	42	48	54
7	0	7	14	21	28	35	42	49	56	63
8	0	8	16	24	32	40	48	56	64	72
9	0	9	18	27	36	45	54	63	72	81

a. Shade the squares that involve multiplication with zero.

Which property allows students to know $0 \times 3 = 0$ and $5 \times 0 = 0$?

b. Shade the squares that involve multiplication with 1.

Which property allows students to know $1 \times 3 = 3$ and $5 \times 1 = 5$?

c. Shade the squares that involve doubles, such as $2 \times 7 = 7 + 7 = 14$.

d. Suppose a student knows $a \times b$. Which property allows a student to quickly determine $b \times a$?

Shade the squares above the main diagonal of the table that allows students to use this property.

e. A *perfect square* is a whole number that can be expressed in the form $n \times n$.

Students often memorize products such as 2×2 and 3×3 because they are perfect squares by

repeating "two times two is four, three times three is nine, …". Shade the squares that are perfect squares.

f. Suppose a student knows 6×6. Discuss how a student can use the distributive property to determine 7×6.

An *oblong number* is a whole number of the form $(n+1) \times n$ or $n \times (n+1)$, such as 7×6 and 8×9.

Shade the squares that are oblong numbers.

g. Students often use skip counting to multiply by five. For example, $3 \times 5 = 15$ because skip counting by five leads

to 5, 10, 15. Shade the squares that involve multiplication by 5.

h. Count the number of unshaded squares in the table. There should be 11 unshaded squares.
If you have fewer or more than 11 unshaded squares, then review parts (a)-(g) and erase or shade squares as needed so that there are 11 unshaded squares. Of the 100 multiplication facts, it appears that 11 of them require some special attention.

i. Devise a strategy to help students multiply by 9. Test your strategy on the products 9×3 and 9×7. Then write your strategy on a piece of paper and ask another student to follow that strategy to determine 9×4 and 9×6.

j. Devise a strategy to help students multiply by 3. Test your strategy on the products 6×3 and 3×8. Then write your strategy on a piece of paper and ask another student to follow that strategy to determine 9×3 and 7×6.

k. Devise a strategy to help students multiply by 4. Test your strategy on the products 6×4 and 7×4. Then write your strategy on a piece of paper and ask another student to follow that strategy to determine 9×4 and 4×8.

ACTIVITY 3.2.1 Fair Share and Repeated Subtraction Models of Division

Material: sharp pencil   Name _____

Multiplication answers the question "How many in all?" when we combine equal-sized groups. Division is the process of separating a collection of objects into equal-sized groups. What two questions does division answer? Division answers the questions *How many groups?* and *How many objects per group?*

1. The **fair share model** of division is appropriate when the total number of objects is known, the number of groups is known, and each group must receive the same number of objects. The fair share model answers the question *How many objects in each group?* Suppose we begin with 12 objects and 3 groups. Then quotient of $12 \div 3$ is the number of objects in each group.

There are 4 objects in each group, so $12 \div 3 = 4$. Draw a diagram to help answer each question.

a. A teacher wants to separate 15 students into 3 teams. How many students are on each team?

b. Vincent has 12 plates to package in boxes. There are 4 boxes. How many plates can he put in each box?

2. Write a word problem for $18 \div 6$ that fits the fair share model of division.

3. Use the diagram to write a division sentence (equation) that answers the question "How many groups?"

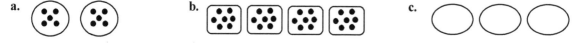

a.   b.   c.

4. The **repeated subtraction model** for division is appropriate when the total number of objects is known, the number of objects in each group is known to be the same, and one wants to determine the number of groups that can be formed. The repeated subtraction model answers the question *How many groups?* Suppose we begin with 12 objects and forms groups with 4 objects. Then quotient of $12 \div 3$ is the number of groups formed.

There are 4 groups, so $12 \div 3 = 4$. Draw a diagram to help answer each question.

a. A teacher wants to separate 15 students into teams of 3. How many teams can be formed?

b. Vincent has 12 plates to package in boxes. Each box can hold 4 plates. How many boxes does he need?

5. Write a word problem for $18 \div 6$ that fits the repeated subtraction model of division.

6. Use the diagram to write a division sentence (equation) that answers the question "How many groups?"

   a.

   b.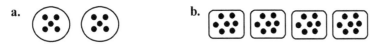

7. Work with a partner.

   a. Use a diagram to help you determine $15 \div 3$. One student should use the fair model of division and the other student should use the repeated subtraction model of division.

   b. Compare the quotients in part (a). Does the quotient depend on the model chosen?

   c. Write a division sentence (equation) for each diagram.

   d. Write a multiplication sentence for each diagram.

   e. Do your equations in parts (c) and (d) suggest that multiplication and division are related?

8. You are given the following diagram.

   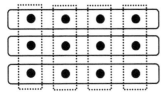

   a. Write an equation involving division.

   b. Write an equation involving multiplication.

9. Write two equations involving multiplication, and two equations involving division for the diagram.

10. Draw a diagram for the equation $14 \div 3 = 4$ R2 that corresponds to the

   a. fair share model.

   b. repeated subtraction model.

11. Solve each equation using the fact that multiplication and division are inverse operations (see Example 3.22 in your textbook).

   a. $n \times 6 = 42$

   b. $k \div 7 = 35$

ACTIVITY 3.2.2 The Quotient-Remainder Theorem

Material: sharp pencil   Name _____

The Quotient-Remainder Theorem: Let a and let b represent any whole numbers such that $b \neq 0$.

There exists unique whole numbers q and r such that $a = b \cdot q + r$, where $0 \leq r < b$.

Note: Another way to express $a = b \cdot q + r$ is $a \div b = q$ Rr, read "a divided by b equals q, with remainder r."

1. Suppose a student writes the given equation. Apply the Quotient-Remainder Theorem to check the student's work.

    a. $34 \div 5 = 6$ R4

    b. $46 \div 6 = 7$ R3

2. Use a calculator to find the quotient q and remainder r for each problem $a \div b$.

    (Hint: Use the calculator to find the quotient q, then define r by $r = a - b \cdot q$.)

    a. $75 \div 18$

    b. $162 \div 19$

    c. $456 \div 32$

3. Mario has 17 books and Kendall has 5 books. From the calculation $17 \div 5 = 3$ R2, we can write "Mario has 2 more than 3 times as many books as Kendall." Use this approach to compare the given quantities.

    a. Lenny has 46 coins and Julia has 14 coins.

    b. There were 12 teachers and 75 students on the field trip.

4. Each problem requires the calculation $30 \div 7 = 4$ R2. Interpret the remainder in each problem.

    a. Marcia has 30 flowers. She wants to plant 7 flowers in each row of a flower bed.
    How many rows would her flower bed have?

    b. 30 students are going on a field trip. Each van can hold at most 7 students.
    How many vans are needed to transport the students?

    c. Kayla has 30 beads. She wants to use all of them to make necklaces. Each necklace requires 7 beads.
    How many more beads would she need?

5. Write a word problem that requires the calculation $22 \div 3 = 7$ R1.
    Pose the question in the word problem in such a way that the answer to the problem is 2.

6. Suppose you know $y \div 5 = q$ R3. Find the quotient and remainder for each problem.

    a. $(y + 2) \div 5$

    b. $(y - 1) \div 5$

    c. $2y \div 5$

ACTIVITY 3.3.1   Making Sense of the Standard Multiplication Algorithm

Material: sharp pencil and base ten blocks (p. A1)   Name _____

You already know how to multiply two multi-digit numbers such as 342 and 43 from your years in elementary school:

```
        1
        ꭓ
      3  4  2
    ×    4  3
  ────────────
      1  0  2  6
  1   3  6  8  0
  ────────────
  1   4  7  0  6
```

The goal of this activity is to help you understand the role of place value in the standard multiplication algorithm.

- **First**, we will use base ten blocks to learn how to multiply a multi-digit number and single digit, such as 64 and 3.

- **Second**, we will use the associative property of multiplication to multiply a power of 10, such as in the problem 342×40.

- **Third**, we will see that a problem such as 342×43 consists of two simpler problems: 342×40 and 342×3, which we already know how to determine. This illustrates that the standard multiplication algorithm is built on simpler problems.

1. This problem uses base ten blocks to find the product: 3×47.

    a. Use base ten blocks to model 47.

    b. Form 3 copies of the model of 47 to model 3×47.

    c. Combine the ones, regrouping as necessary.

    d. Combine the tens, regrouping as necessary.

    e. Use the base ten blocks to determine 3×47

2. Let's analyze the standard algorithm for determining 3×47.

```
      1   2
        4  7
    ×      3
  ──────────
    1   4  1
```

    To answer the following questions, think about how you would combine base ten blocks to find the product.

    a. A student asks why the 2 appears over the 4. Fill in the blanks to complete the following reply:

    "3 times 7 ones = _____ ones. The 2 over the 4 represents the 2 tens from regrouping _____ ones as

    _____ tens, _____ ones."

    b. A student asks why the 1 appears in the hundreds place. Fill in the blanks to complete the following reply:

    "3 times 4 tens + 2 tens = 12 tens + 2 tens = _____ tens. The 1 in the hundreds place represents the

    1 hundred from regrouping 14 _____ as _____ hundred, _____ tens."

3. Let's analyze the standard multiplication algorithm for determining 4×68.

    a. Explain why the 3 appears over the 6.

    b. Explain why the 2 appears in the hundreds place.

$$
\begin{array}{c}
\ {}^{2}\ {}^{3}\\
\ 6\ 8\\
\times \ 4\\
\hline
2\ 7\ 2
\end{array}
$$

4. Some examples of powers of 10 are: 10, 100, and 1000. Some examples of multiples of a power of 10 are: 670, 6700, and 67,000. We can use the associative property to make it easier to multiply by a multiple of a power of 10. For example, $3 \times 5400 = 3 \times (54 \times 100) = (3 \times 54) \times 100$. Then we would just need to find 3×54 : $3 \times 54 = 162$, so $3 \times 5400 = 162 \times 100 = 16,200$. Fill in the blank with the appropriate power of 10.

    a. $4 \times 3800 = (4 \times 38) \times \underline{\hspace{2cm}}$

    b. $7 \times 3290 = (7 \times 329) \times \underline{\hspace{2cm}}$

    c. $300 \times 42 = (3 \times 42) \times \underline{\hspace{2cm}}$

    d. $60 \times 75000 = (6 \times 75) \times \underline{\hspace{2cm}}$

5. Let's consider the problem 243×67. Expand the factor 67 and use the distributive property: $243 \times 67 = 243 \times (60 + 7) = 243 \times 60 + 243 \times 7$. We see that the product 243×67 is the sum of two simper problems: 243×7 and 243×60. We already know how to solve each one of these simpler problems: $243 \times 7 = 1701$ and $243 \times 60 = (243 \times 6) \times 10 = 14580$. The sequence of computations are given below:

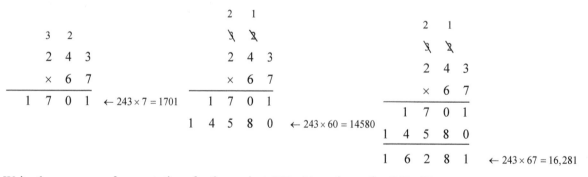

Write the sequence of computations for the product 263×54 as shown for 243×67.

6. Write two simpler problems that 742×59 requires using the standard multiplication algorithm.

7. Write three simpler problems that 628×843 requires using the standard multiplication algorithm.

8. Mike is using the standard multiplication algorithm to calculate 547×83, as shown.

    Why did he write a 0 in the ones place?

    ```
        1   2
        5   4   7
      ×     8   3
    ———————————————
      1   6   4   1
                  0
    ```

9. Desiree is using the standard multiplication algorithm to calculate 742×578, as shown.

    Why did she write a 0 in the ones and tens places?

    ```
      2   1
      X   X
      2   7   4
    × 6   3   8
    ———————————————
    2   1   9   2
    8   2   2   0
              0   0
    ```

10. You are given that $abc \times d = 1728$ and $abc \times e = 2304$.

    a. What is $abc \times de$?

    b. What is $abc \times ed$?

11. You are given that $rst \times u = 3416$ and $rst \times v = 2135$.

    a. What is $rst \times uv$?

    b. What is $rst \times vu$?

ACTIVITY 3.3.2   Making Sense of Long Division

Material: sharp pencil and base ten blocks (p. A1)   Name _____

The standard division algorithm with whole numbers, also called long division, is straightforward to illustrate using the fair share model of division ("How many in each group?"). To keep ideas simple, we use a single-digit divisor. The dividend represents the total number of objects, the divisor represents the number of groups, and the quotient (which is to be determined) represents the number of objects in each group. There may be a remainder. The base ten blocks provide a solid foundation for making sense of long division because they encourage logical reasoning with place value concepts.

Example: Use base ten blocks to divide using the fair share model: $3\overline{)43}$.

Step 1. Build the dividend 43 using base ten blocks.

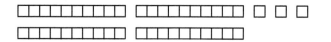

Step 2. We need to divide the tens equally among the 3 groups.

Each gets 1 ten. 4 tens – 3 tens = 1 ten. There are 1 ten and 3 ones remaining.

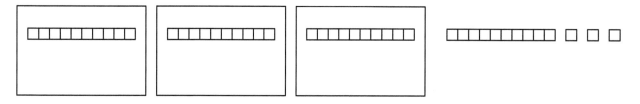

Step 3. Regroup the 1 ten as 10 ones. Then 10 ones + 3 ones = 13 ones.

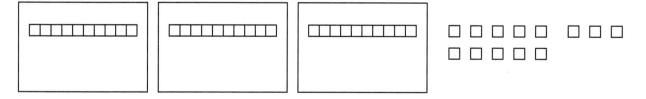

Step 4. Now we need to give away the ones. Divide the ones equally among the 3 groups.

Each group gets 4 ones. 13 ones – 12 ones = 1 one. Then 1 one remains.

Step 5. Each group gets 1 ten and 4 ones, with 1 one leftover.

Then $43 \div 3 = 14$ R1 .

Now we are ready to describe the steps of the long division. Follow along with the base ten blocks.

Step 1. Set up the problem.

$$3\overline{)4\ 3}$$

Step 2. Divide 4 tens into 3 equal-sized groups. Each group get 1 ten ("put the 1 above the 4").
4 tens − 3 tens = 1 ten. That leaves 1 ten and 3 ones.

$$\begin{array}{r} 1\ 4 \\ 3\overline{)4\ 3} \\ \underline{-3} \\ 1 \end{array}$$

Step 3. Regroup the 1 ten and 3 ones as 13 ones ("bring down the 3").
Note: The conceptual basis for bringing down a digit is *regrouping*.

$$\begin{array}{r} 1\ 4 \\ 3\overline{)4\ 3} \\ \underline{-3\downarrow} \\ 1\ 3 \end{array}$$

Step 4. Divide 13 ones into 3 equal-sized groups. Each group get 4 ones ("put the 4 above the 3").
13 ones − 12 ones = 1 one. Then 1 one remains.

$$\begin{array}{r} 1\ 4 \\ 3\overline{)4\ 3} \\ \underline{-3\downarrow} \\ 1\ 3 \\ \underline{-1\ 2} \\ 1 \end{array}$$

Step 5. Then $43 \div 3 = 14$ R1.

---

1. Work in pairs. Use the base ten blocks to divide. Discuss the regroupings that occur. At each step, write the corresponding step in the pencil-and-paper long division algorithm to make a connection between the base ten blocks and long division. Discuss what is going on when you "bring down" a digit.

    a. $4\overline{)76}$ **b.** $3\overline{)53}$ **c.** $3\overline{)475}$

2. A student uses long division to divide: $957 \div 4$. Part of the work is shown.
    Answer the following questions. Compare your answers with another student.

    $$\begin{array}{r} 2\ 3 \\ 4\overline{)9\ 5\ 7} \\ \underline{-8\downarrow} \\ 1\ 5 \\ \underline{-1\ 2\downarrow} \\ 3\ 7 \end{array}$$

    a. Explain why the 2 in the quotient appears above the 9 in the dividend.

    b. Explain why we subtract 8.

    c. Describe the regrouping that occurs when we bring down the 5.

    d. Explain why the 3 in the quotient appears above the 5 in the dividend.

    e. Explain why we subtract 12.

    f. Describe the regrouping that occurs when we bring down the 7.

3. Create your own problem similar Problem 2. Give the problem to another student. Check their answers.

ACTIVITY 3.3.3   Multiplication by Duplation (Successive Doubling)

Material: sharp pencil   Name _____

Ancient Egyptians devised a simple, but clever process called **duplation** (successive doubling) to multiply numbers. Duplation exploits the fact that every whole number can be written as a sum of distinct powers of 2: 1, 2, 4, 8, 16, 32, 64, and so on. For example: $12 = 8 + 4$, $53 = 32 + 16 + 4 + 1$, and $73 = 64 + 8 + 1$. Duplation also exploits the fact that numbers are easy to double by adding: $15 + 15 = 30$, $46 + 46 = 92$, and so on. Although ancient Egyptians used hieroglyphics (pictures such as the coiled rope) in their number system, we demonstrate this historical and cultural connection using base ten numbers to make duplation easier to understand.

Let's use duplation to find the product $25 \cdot 63$. Duplation involves composing the multiplier (25) from appropriate powers of 2 (1, 2, 4, 8, 16, 32, ...). We will organize the results in a table.

Step 1. The left column contains powers of 2. List the powers of 2 up to the multiplier 25.

The entry 32 exceeds the multiplier 25, so we cross out the 32 and stop entering powers of 2.

$25 \cdot 63$	
1	
2	
4	
8	
16	
~~32~~	

Step 2. The right column begins with 63, followed by doubles: $63 + 63 = 126$, $126 + 126 = 252$, and so on.

$25 \cdot 63$			
1	63		*Think: 1 group of 63 is 63*
2	126	*Do: 63 + 63 = 126*	*Think: 2 groups of 63 is 126*
4	252	*Do: 126 + 126 = 252*	*Think: 4 groups of 63 is 252*
8	504	*Do: 252 + 252 = 504*	*Think: 8 groups of 63 is 504*
16	1008	*Do: 504 + 504 = 1,008*	*Think: 16 groups of 63 is 1,008*
~~32~~			

Step 3. We use the numbers in the left column to build the multiplier 25. We begin with the largest number (16). Subtracting, we get $25 - 16 = 9$, so we can add the next largest row with 8. Subtracting, we get $9 - 8 = 1$, so we cross out the 4 and 2, and add the first row with 1. Then $25 = 16 + 8 + 1$.

~~$25 \cdot 63$~~	
1	63
2	126
4	252
8	504
16	1008

Here's why it works:
$$
\begin{aligned}
25 \cdot 63 &= (16 + 8 + 1) \cdot 63 \\
&= 16 \cdot 63 + 8 \cdot 63 + 1 \cdot 63 \\
&= 1008 + 504 + 63 \\
&= 1575
\end{aligned}
$$

Step 4. Add the numbers in the uncrossed rows in the right column to find the product: $25 \cdot 63 = 1008 + 504 + 63 = 1575$.

1.   A student uses duplation to multiply: $50 \cdot 35$. Their work is shown below. Find the product.

50 • 35	
~~1~~	~~35~~
2	70
~~4~~	~~140~~
~~8~~	~~280~~
16	560
32	1120
~~64~~	~~2240~~

2.   A student uses duplation to multiply: $26 \cdot 31$. Their partially completed table is shown below.

Cross out the appropriate rows, then use the table to determine the product.

26 • 31	
1	31
2	62
4	124
8	248
16	496
32	992

3.   A student uses duplation to multiply: $51 \cdot 56$. Their partially completed table is shown below.

Cross out the appropriate rows, then use the table to determine the product.

51 • 56	
1	56
2	112
4	224
8	448
16	896
32	1792
64	3584

4.   Use duplation to multiply. Check your work with another student. Your tables should match.

You should also calculate the same product.

   a. $15 \cdot 36$

   b. $58 \cdot 45$

   c. $126 \cdot 42$

   d. $164 \cdot 241$

ACTIVITY 3.3.4   Division by Duplation (Successive Doubling)

Material: sharp pencil   Name _____

The ancient Egyptians used duplation (successive doubling) to divide, much in the same way for multiplication. We will illustrate this method with an example. Let's calculate $143 \div 15$. We know $143 \div 15 = q \text{ R} r$ if and only if $143 = q \cdot 15 + r$. Our goal is to determine how many groups (q) of 15 are in 143. The entries in the left column are multiples of 2. The entries in the right column are multiples of 15 obtained by doubling (15, 30, 60, 120, …). The next entry in the right column of the table would be 240, which exceeds the dividend 143, so we cross out the last row.

$143 \div 15$	
1	15
2	30
4	60
8	120
~~16~~	~~240~~

Think: 1 group of 15 is 15
Think: 2 groups of 15 is 30
Think: 4 groups of 15 is 60
Think: 8 groups of 15 is 120
Think: 16 groups of 15 is 240, which is larger than the dividend 143.

Now we use the numbers in the right column to build the dividend 143.

Step 1.   We begin with the highest value, 120. Then we add the next highest value ($120 + 60 = 180$);
the sum 180 exceeds the dividend 143, so we cross out that row.

$143 \div 15$	
1	15
2	30
~~4~~	~~60~~
8	120
~~16~~	~~240~~

Step 2.   Then we add the next highest value ($120 + 30 = 150$); the sum 150 exceeds the dividend 143, so we cross out that row. Then we add the next highest value ($120 + 15 = 135$); the sum does not exceed the dividend 143. There are $8 + 1 = 9$ groups of 15 in 143, and $9 \cdot 15 = 135$. The remainder is $r = 143 - 135 = 8$. Then $143 = 9 \cdot 15 + 8$, so $143 \div 15 = 9 \text{ R} 8$.

~~$143 \div 15$~~	
1	15
2	30
4	60
8	120
16	240

> *Here's why it works:*
> We decomposed the dividend in terms of a multiple of 15:
> $143 = 9 \cdot 15 + 8$. By the Quotient-Remainder Theorem,
> it follows $143 \div 15 = 9 \text{ R} 8$.

1. A student uses duplation to divide: $600 \div 32$. Their work is shown below.

    Discuss why the rows with 32, 8, 4, and 1 are crossed out.

$600 \div 32$	
~~1~~	~~32~~
2	64
~~4~~	~~128~~
~~8~~	~~256~~
16	512
~~32~~	~~1024~~

    $600 - 512 = 88$, $88 - 64 = 24$, and 24 is less than the divisor 32. Then the remainder is 24.

    Then $600 = 512 + 88 = 512 + 64 + 24 = 16 \cdot 32 + 2 \cdot 32 + 24 = 18 \cdot 32 + 24$. Then $600 \div 32 = 18 \text{ R} 24$.

2. A student uses duplation to divide: $370 \div 18$. Their partially completed table is shown below.

    Cross out the appropriate rows, then use the table to determine the quotient and remainder.

$370 \div 18$	
1	18
2	36
4	72
8	144
16	288
32	576

3. A student uses duplation to divide: $6172 \div 70$. Their partially completed table is shown below.

    Cross out the appropriate rows, then use the table to determine the quotient and remainder.

$6172 \div 70$	
1	70
2	140
4	280
8	560
16	1120
32	2240
64	4480
128	8960

4. Use duplation to divide. Check your work with another student. Your tables should match.

    You should also calculate the same quotient and remainder.

    a. $197 \div 17$

    b. $875 \div 41$

    c. $421 \div 23$

    d. $317 \div 14$

ACTIVITY 4.1.1   Remember the Divisibility Rules!

Material: sharp pencil   Name _____

The goal of this activity is to reinforce the basic divisibility rules.

- 2 divides the whole number *n* if and only if the ones digit of *n* is 0, 2, 4, 6, or 8.
- 3 divides the whole number *n* if and only if 3 divides the sum of the digits of *n*.
- 4 divides the whole number *n* if and only if 4 divides the number formed by the two rightmost digits of *n*.
- 5 divides the whole number *n* if and only if the ones digit of *n* is 0 or 5.
- 6 divides *n* if and only if 2 divides *n* and 3 divides *n*.
- 8 divides the whole number *n* if and only if 8 divides the number formed by the three rightmost digits of *n*.
- 9 divides the whole number *n* if and only if 9 divides the sum of the digits of *n*.
- 10 divides the whole number *n* if and only if the ones digit of *n* is 0.

Game. This game involves a group of four students. Each student secretly writes one digit (0, 1, 2, ..., 9) on a piece of paper. Then the students display their digit to the whole group so that four digits are shown. Each student secretly writes a four-digit number composed from the four digits on a piece of paper. Then each student applies the divisibility rules and writes the numbers (2, 3, 4, 5, 6, 8, 9, or 10) that divide their four-digit number (note: calculators are not allowed in this game). Then the students share their results with the students in the group to check their answers. Each student should follow these steps to determine their own score: one point is awarded for each divisibility rule that is correctly identified, and one point is subtracted for each divisibility rule that is incorrectly identified, and one point is subtracted for each applicable divisibility rule that was not identified. For example, suppose a student composed the four-digit number 4614 and listed its divisors as 2, 3, 6, and 9. The correct divisors of 4614 are 2, 3, and 6. Then the student earns 3 points for identifying three correct divisibility rules but loses one point for incorrectly identifying a divisibility rule, for a total of 3 − 1 = 2 points. The student with the highest score earns 3 points. The student with the second highest score earns 2 points. The person with the third highest score earns 1 point. If there is a two-way tie for first place, then the student with the lowest four-digit number would earn 3 points and the other student would earn 2 points. If there is a two-way tie for second place, then the student with the lowest four-digit number would earn 2 points and the other student would earn 1 point, and so on. The tie-breaking rules can be adapted for other possible situations. Each game consists of four timed rounds to select a 4-digit number. The time for a round begins as soon as the four digits are shared with the group so that time ticks away while students choose their four-digit number. Round 1 lasts 25 seconds. Round 2 lasts 20 seconds. Round 3 lasts 15 seconds. Round 4 lasts 10 seconds. The student with the highest total of points in their group is the winner. Note: Leading zeros in a four-digit number are ignored For example, the four-digit numbers 0561 and 0052 should be treated as 561 and 52, respectively, when the divisibility rules are applied.

Variations: This game can be played with more students. Also, the students can decide how ties should be broken.

ACTIVITY 4.1.2   Extending Some Divisibility Rules

Material: sharp pencil   Name _____

Number theory is the study of properties of counting numbers, their relationships, and ways to represent them. Number theory concepts in the elementary mathematics curriculum involve factors, multiples, divisors, divisibility rules, prime factorization, least common multiple (LCM), and greatest common factor (GCF). Together, these ideas form a foundation for understanding how counting numbers are related through multiplication and division.

A divisibility rule is a simple procedure to check if one number divides another number, without performing the original division. Here's some vocabulary that we will need: Let a and b represent whole numbers with $b \neq 0$. If a whole number q exists such that $a = b \cdot q$ then we say:

- b is a **factor** of a
- b is a **divisor** of a
- b **divides** a
- a is a **multiple** of b
- a is **divisible** by b

Section 4.1 of the text discusses many different divisibility rules. This activity will be exploring some extensions.

1. The divisibility rule for the number 2 is "2 divides the whole number n if and only if the ones digit of n is 0, 2, 4, 6, or 8." The divisibility rule for the number 4 is "4 divides the whole number n if and only if 4 divides the number formed by the two rightmost digits of n." The divisibility the rule for the number 8 is "8 divides the whole number n, if and only if, 8 divides the number formed by the three rightmost digits of n." Use a divisibility rule for each problem. Tell how you arrived at your answer.

    a. Does 2 divide 35,670,567?

    b. Does 2 divide 25,328,486?

    c. Does 4 divide 39,020?

    d. Does 4 divide 456,762?

    e. Does 8 divide 345,641,984?

    f. Does 8 divide 2,450,394?

2. Do the following.

    a. Find the largest digit a such that 2 is a divisor of 745a4.

    b. Find the largest digit b such that 4 is a factor of 124,5b6.

    c. Find the largest digit c such that 8 divides 3,789,63c.

3. Apply divisibility rules to explain why

   a. there is no whole number a such that solves the equation $8a = 456{,}789{,}123{,}310$.

   b. there are no whole numbers a and b that solve the equation $20a + 36b = 128{,}122$.

4. 2, 4, and 8 are powers of 2. What is the next power of 2?

5. Discuss the pattern in the divisibility rules for 2, 4, and 8.

6. Extend that pattern by stating a divisibility rule for the number 16.

7. According to your divisibility rule, does 16 divide 239,762,592?

8. Using a calculator, divide 239,762,560 by 16. Does 16 divide 239,762,592?

9. Generalize your results by stating a divisibility rule for the number 2^n , where n is a counting number.

The following group of questions relate to divisibility rules for 3 and 9.

10. State the divisibility rule for the number 3.

11. Is 6735 divisible by 3? Tell how you know.

12. State the divisibility rule for the number 9.

13. Is 4267 divisible by 9? Tell how you know.

14. Apply divisibility rules to explain why

   a. there is no whole number a such that solves the equation $3a = 67{,}452{,}118$.

   b. there are no whole numbers a and b that solve the equation $12a + 30b = 456{,}781$.

ACTIVITY 4.1.3   ISBN Numbers and Divisibility

Material: sharp pencil   Name _____

ISBN is an acronym for International Standard Book Number. An ISBN is a 13-digit number that uniquely identifies a book for marketing purposes. The first twelve (leftmost) digits encode the geographical location of the publisher, the language of the book, the publisher, and the title of the work. The thirteenth (rightmost) digit is called the *checksum digit*. The checksum digit is designed to detect an error when an ISBN is scanned or typed in a computer. For example, the book *Notable Women in Mathematics: A Biographical Dictionary* has the ISBN 978-0-313-29131-9. The following table shows the weights for each digit in any ISBN, and illustrates how to calculate the weighted sum for the given ISBN 978-0-313-29131-9. The weights are always 1, 3, 1, 3, 1, 3, 1, 3, 1, 3, 1, 3, and 1.

ISBN	9	7	8	0	3	1	3	2	9	1	3	1	9
weight	1	3	1	3	1	3	1	3	1	3	1	3	1
weighted sum	$9 \cdot 1 + 7 \cdot 3 + 8 \cdot 1 + 0 \cdot 3 + 3 \cdot 1 + 1 \cdot 3 + 3 \cdot 1 + 2 \cdot 3 + 9 \cdot 1 + 1 \cdot 3 + 3 \cdot 1 + 1 \cdot 3 + 9 \cdot 1 = 80$												

A 13-digit number is a valid ISBN provided the weighted sum is divisible by 10. The weighted sum of 978-0-313-29131-4 is 80. Because 80 is divisible by 10, we conclude that 978-0-313-29131-4 is a *valid* ISBN number. Suppose you wrote 978-0-313-59131-4 because you wrote a 5 instead of a 2 in the ISBN number. The weighted sum would be $9 \cdot 1 + 7 \cdot 3 + 8 \cdot 1 + 0 \cdot 3 + 3 \cdot 1 + 1 \cdot 3 + 3 \cdot 1 + 5 \cdot 3 + 9 \cdot 1 + 1 \cdot 3 + 3 \cdot 1 + 1 \cdot 3 + 9 \cdot 1 = 89$. The weighted sum is 89, which is not divisible by 10, so 978-0-313-59131-4 is an *invalid* ISBN number.

1. The classic book *A History of Mathematics (Second Ed.)* by Boyer and Merzbach with ISBN 978-0-471-54397-8 provides a comprehensive, chronological, and accessible summary of the historical development of mathematics through the ages. Verify that the 13-digit number is a valid ISBN.

2. Determine if the ISBN is valid or invalid.
    a. 978-5-732-17868-7
    b. 978-3-628-59277-5.

3. Find a digit *d* such that 978-4-723-3*d*542-1 is a valid ISBN.

4. The ISBN number for the book *A History of Mathematical Notations* is 978-0-486-67766-*n*. Determine the checksum digit *n*.

5. The checksum digit is the last of the thirteen digits to be assigned, and is based on a weighted sum of the first twelve digits. The weighted sum of the twelve leftmost digits of an ISBN is given. What is the checksum digit?
    a. 128   b. 114   c. 130

6. The ISBN for a book has 13 digits. An ISBN is valid as long as the weighted sum of the digits is a multiple of 10. Would the method be more or less prone to errors if the weighted sum had to be a multiple of 3 instead of a multiple of 10? Explain.

ACTIVITY 4.2.1 Listing and Counting the Factors of a Number

Material: sharp pencil   Name _____

A *composite number* is a counting number that has more than two factors. A *prime number* is a counting number that has exactly two factors. The number 1 is neither a composite nor a prime number. The equation $24 = 2 \cdot 3 \cdot 4$ is an example of a *factorization*. The equation $24 = 2 \cdot 2 \cdot 2 \cdot 3$ is an example of a *prime factorization*, which can be expressed using exponential notation as $24 = 2^3 \cdot 3$. According to the Fundamental Theorem of Arithmetic, all composite numbers can be expressed as a unique product of primes (as long as we ignore the ordering of the factors). Letters of the alphabet are building blocks of words. Similarly, prime numbers are the building blocks of numbers. In this activity, we will learn how to methodically list the factors of a composite number, and learn how to use the prime factorization of a number to determine its number of factors, number of prime factors, and number of composite factors.

1. The diagram shows two rectangular array models of 12. Discuss how they indicate that 12 is a composite number.

    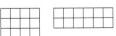

2. List all the factors of 24. Discuss the approach you used to determine all of the factors.

3. A factor pair of 24 is a pair of numbers a and b such that $24 = a \cdot b$. One factor pair is 1 and 24. List all the factor pairs of 24.

4. The diagram shows all the factors of 24 and 36 in increasing order. What patterns did you notice?

    1   2   3   4   6   8   12   24   1   2   3   4   6   9   12   18   36

5. Use the following strategy, called **repeated division**, to make a list of all the factors of the composite number a. 1 is a factor, so put 1 and a in the list. Then divide a by 2. If 2 divides a, then add 2 and $a \div 2$ to the list. Then divide a by 3. If 3 divides a, then add 3 and $a \div 3$ to the list. Repeat this process for divisors 4, 5, 6, ... up to a. Once a factor in the list reappears, then the list contains all of the factors of a, and the procedure terminates. Use repeated division to list all the factors of 24. Write the factors in the order they are discovered.

6. Use repeated division to list all the factors of 36. Write the factors in the order they are discovered.

7. Use repeated division to list all the factors of 42. Write the factors in the order they are discovered.

8. The prime factorization of 24 is $24 = 2^3 \cdot 3$. Each factor of 24 can be expressed as a factorization using exponential notation. Write the appropriate exponent for each factor of 24.

$1 = 2^\square \cdot 3^\square$	$2 = 2^\square \cdot 3^\square$	$3 = 2^\square \cdot 3^\square$	$4 = 2^\square \cdot 3^\square$
$6 = 2^\square \cdot 3^\square$	$8 = 2^\square \cdot 3^\square$	$12 = 2^\square \cdot 3^\square$	$24 = 2^\square \cdot 3^\square$

9. 567 has the prime factorization $567 = 3^4 \cdot 7^1$, and 567 has 10 factors.

    1375 has the prime factorization $1375 = 5^3 \cdot 11^1$, and 1375 has 8 factors.

    10,976 has the prime factorization $10,976 = 2^5 \cdot 7^3$, and 10,976 has 24 factors.

    a. Look for a pattern in the examples to devise a method for determining the number of factors of a number based on the prime factorization of the number.

    b. Fill in the blank: 24 has _____ factors.

    c. Express the prime factorization of 24 using exponential notation.

    d. Using the prime factorization in part (c), apply your strategy from part (a) to predict the number of factors of 24.

        Your answer: 24 has _____ factors.

    e. Do the answers in part (b) and (d) match? If not, then please revise your strategy in part (a) or revise your answer in part (b), or both, and then answer this question again.

10. The following set contains every factor of 392: {1, 2, 4, 7, 8, 14, 28, 49, 56, 98, 196, 392}.

    a. Determine the number of prime factors of 392.

    b. Determine the number of composite factors of 392.

11. The prime factorization of 392 is $392 = 2^3 \cdot 7^2$.

    a. How can you determine the number of factors of 392 from the prime factorization of 392?

    b. How can you determine the number of prime factors of 392 from the prime factorization of 392?

    c. How can you calculate the number of composite factors of 392 from the prime factorization of 392?

12. n has the prime factorization $n = 2^3 \cdot 7^5 \cdot 11^9$ and has 240 factors. How can you use the prime factorization of n to determine another number that has 240 factors?

13. Do the following.

    a. Draw a prime factor tree for 90.

    b. Use the tree diagram to find the prime factorization of 90.

    c. Determine the number of factors of 90.

    d. How many factors of 90 are prime numbers?

    e. How many factors of 90 are composite numbers?

14. Do the following.

    a. How many factors does $3^6 \cdot 5^8 \cdot 11^2$ have?

    b. How many prime factors does $3^6 \cdot 5^8 \cdot 11^2$ have?

    c. How many composite factors does $3^6 \cdot 5^8 \cdot 11^2$ have?

ACTIVITY 4.2.2 Finding Prime Numbers Using the Sieve

Material: sharp pencil   Name _____

Eratosthenes, a Greek mathematician in the third century BCE, proposed a simple procedure called the "Sieve of Eratosthenes" for finding all the prime numbers in the list 1, 2, 3, …, N. A sieve is like a strainer that separates wanted elements (prime numbers) from unwanted elements (composite numbers) in an orderly fashion. It is an efficient procedure for finding prime numbers less than 10,000,000,000 (http://primes.utm.edu). Now we will lead you through this procedure for finding all the prime numbers from 1 to 100. The prime numbers will be the circled numbers and the composite numbers will be shaded.

1	2	3	4	5	6	7	8	9	10
11	12	13	14	15	16	17	18	19	20
21	22	23	24	25	26	27	28	29	30
31	32	33	34	35	36	37	38	39	40
41	42	43	44	45	46	47	48	49	50
51	52	53	54	55	56	57	58	59	60
61	62	63	64	65	66	67	68	69	70
71	72	73	74	75	76	77	78	79	80
81	82	83	84	85	86	87	88	89	90
91	92	93	94	95	96	97	98	99	100

1. Shade the cell with 1 because it is neither a prime nor composite number.

2. The first prime number is 2, so circle the number 2. Then shade the boxes containing multiples of 2.

3. After you circled 2, what is the first number you shaded?

4. The next prime number is 3, so circle the number 3. Then shade the boxes containing multiples of 3. After you circled 3, what is the first number you shaded?

5. The first number that you shade when you shade the multiples of 3 is not 6. Why?

6. The next prime number is not 4. Use the shading of the grid to answer why.

7. The next prime number is 5, so circle the number 5. Then shade the boxes containing multiples of 5. After you circled 5, what is the first number you shaded?

8. The next prime number is 7, so circle the number 7. Then shade the boxes containing multiples of 7. After you circled 7, what is the first number you shaded?

9. What is the next unshaded number?

10. The next unshaded number should be 11 (you may need to fix your sieve). Circle the number 11.

If you were to shade the multiples of 11, what would be the first number you would shade and why?

11. Since 121 is the next number you would shade and is beyond the upper limit of the list, all of the remaining unshaded squares are prime numbers. Circle all of the unshaded numbers.

12. All of the circled numbers are prime numbers. List all of the prime numbers that you discovered. Does your list match the list of primes obtained by another student?

13. What is the first number that you would shade when shading multiples of any prime number p? Justify your answer.

ACTIVITY 4.2.3 The Least Common Multiple and the Greatest Common Factor

Material: sharp pencil   Name _____

We will explore the prime factorization method for finding the least common multiple (LCM) and the greatest common factor (GCF) of two numbers, and then explore the relationship between the LCM and GCF of two numbers. The symbol LCM(a, b) represents the least common multiple of a and b, and the symbol GCF(a, b) represents the greatest common factor of a and b.

1. Multiples of 12 are 12, 24, 36, 48, 60, 72, ... Multiples of 18 are 18, 36, 54, 72, ...
    a. Find the next three terms of each sequence.

    b. List the first three common multiples of 12 and 18.

    c. What is the least common multiple of 12 and 18?

2. The factors of 12 are 1, 2, 3, 4, 6, and 12. The factors of 18 are 1, 2, 3, 6, 9, and 18.
    a. List the common factors of 12 and 18.

    b. What is the greatest common factor of 12 and 18?

3. The prime factorizations of 40 and 300 are $40 = 2^3 \cdot 5$ and $300 = 2^2 \cdot 3 \cdot 5^2$.
    a. LCM(40, 300) = 600, which has the prime factorization LCM(40, 300) = $2^3 \cdot 3 \cdot 5^2$. Discuss how the prime factorization of the LCM of 40 and 300 can be derived from the prime factorizations of 40 and 300.

    b. GCF(40, 300) = 20, which has the prime factorization GCF(40, 300) = $2^2 \cdot 5$. Discuss how the prime factorization of the GCF of 40 and 300 can be derived from the prime factorizations of 40 and 300.

4. In the previous problem, you discussed how to use prime factorizations to find the LCM and GCF. The purpose of this problem is to check your thinking. The prime factorizations of a and b are $a = 2^3 \cdot 5 \cdot 7^4$ and $b = 2^6 \cdot 3 \cdot 7^2$.
    a. Use your approach in Problem 3(a) to find LCM(a, b). Write the answer in prime factorization form.

    b. Use your approach in Problem 3(b) to find GCF(a, b). Write the answer in prime factorization form.

    c. Now check your answers: LCM(a, b) = $2^6 \cdot 3 \cdot 5 \cdot 7^4$ and GCF(a, b) = $2^3 \cdot 7^2$.

5. Suppose $a = 2^5 \cdot 3^2 \cdot 7^4$ and $b = 2^5 \cdot 3 \cdot 5^2 \cdot 7^2$. Use these prime factorizations to find each number. Write the answer in prime factorization form.

    a. LCM(a, b)

    b. GCF(a, b)

6. Use the prime factorization method to find each number.

    a. LCM(60, 72)

    b. GCF(60, 72).

7. Do the following.

    a. Complete the table.

a	b	$a \cdot b$	GCF(a, b)	LCM(a, b)	GCF(a, b) $\cdot$ LCM(a, b)
4	6		2	12	
3	5		1	15	
32	48		16	96	

    b. What relationship do you see between $a \cdot b$ and GCF(a, b) $\cdot$ LCM(a, b)?

    c. Use the fact that LCM(128, 240) = 1920 to find GCF(128, 240).

    d. Use the fact that GCF(672, 540) = 12 to find LCM(672, 540).

8. A manufacturer puts a coupon worth $1.00 in every 15th box of cereal and a prize in every 18th box of cereal. If you buy 1000 boxes of cereal, how many boxes would you expect to contain both the coupon and prize?

9. There are 72 boys and 60 girls in a summer camp. The counselor wants to split them into equal-sized groups such that each group has the same number of boys and each group has the same number of girls.

    a. What is the maximum number of groups that can be formed?

    b. For your answer in part (a), how many people would belong to each group?

ACTIVITY 4.2.4 The Euclidean Algorithm

Material: sharp pencil   Name _____

In this activity, we will focus on how to use *division* (rather than the prime factorization) to find the greatest common factor of two numbers. This method is called the *Euclidean Algorithm*. We will demonstrate the Euclidean Algorithm with a simple example.

Find the GCF of 56 and 188. We can relate these numbers through division: $188 \div 56 = 3$ R20. Then $188 = 3 \cdot 56 + 20$, and any number d that divides 56 and 188 must also divide the difference $188 - 3 \cdot 56$ (which is 20). Do you agree that d must divide 20, too? Please think about this before reading further. If you agree, then the problem of finding the GCF of 56 and 188 is equivalent to the simpler problem of finding the GCF of 56 and 20. *The Euclidean Algorithm replaces the larger number with the remainder to obtain a simpler problem.* Then divide: $56 \div 20 = 2$ R16, or $56 = 2 \cdot 20 + 16$, and replace the larger number with the new remainder to obtain the simpler problem of finding the GCF of 16 and 20. Then divide: $20 = 1 \cdot 16 + 4$, and replace the larger number with the new remainder to obtain the simpler problem of finding the GCF of 16 and 4. Then divide: $16 = 4 \cdot 4 + 0$, and replace the larger number with the new remainder to obtain the simpler problem of finding the GCF of 0 and 4. Then the GCF is 4, since 4 divides both 0 and 4. The process can be summarized with a sequence of equations:

Euclidean Algorithm

GCF(56, 188)	= GCF(56, 20)	because $188 \div 56 = 3$ R20
	= GCF(16, 20)	because $56 \div 20 = 2$ R16
	= GCF(16, 4)	because $20 \div 16 = 1$ R4
	= GCF(0, 4)	because $16 \div 4 = 4$ R0
	= 4	because 4 divides 0 and 4

The divisions stop when the remainder is 0 or 1 because $GCF(0, n) = n$ and $GCF(1, n) = 1$.

1. Use the Euclidean Algorithm to find the GCF of each pair of numbers. Summarize the calculations as shown above. Compare your equations with another student in the class. You should have the same equations.

    a. 475 and 90

    b. 270 and 396

    c. 128 and 1880

2. How could you use the Euclidean Algorithm to help you find the LCM of a and b? (*Hint*: Remember how the products $a \cdot b$ and $GCF(a, b) \cdot LCM(a, b)$ are related.)

3. Use the Euclidean Algorithm to find the LCM of each pair of numbers.

    a. 72 and 135

    b. 84 and 280

    c. 234 and 567

ACTIVITY 4.3.1   Adding and Subtracting Integers Using the Chip Model

Material: sharp pencil and two color chips (page A2)   Name _____

Adding and subtracting integers is a critical arithmetic and algebraic milestone for students. This activity explores adding and subtracting integers using the chip model. This model provides a concrete way for students to practice, develop, and internalize rules for adding and subtracting with integers. In this activity, the two color chip model uses black and white chips, although any two colors would work, such as red and yellow. The chips may also be called *counters*.

Chips provide memorable experience for students to physically represent the symbolic nature of integers. An important rule with respect to the two color chips is the *zero pair rule*: the combination of one white chip and one black chip equals zero. Below are some examples. Note that one white chip represents 1, and one black chip represents -1.

chips:	O	●	O●	O O	●●	O O O	●●●	O O●	O O●●●
integer:	1	-1	0	2	-2	3	-3	1	-1

A **drop** of two color chips involves taking a handful of chips and dropping them on a table. To simplify a drop of chips, use the zero pair rule until there are only chips of one color left on the table. Then the integer is determined by the color and number of chips. Figure 1 is the initial drop of colored chips. In Figure 2, we circle zero pairs. There are two black chips remaining. So the resulting integer is -2.

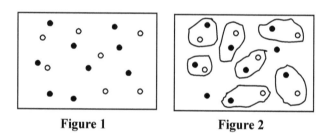

Figure 1   **Figure 2**

Addition of Integers using the two color chip model. To add two integers, first model both integers on the table. Second, use the zero pair rule to simplify the representation on the table. The sum is the integer represented by the remaining chips. For example, let's use the chip model to add: $-3+5$.

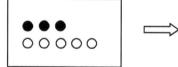

Model each addend.   Use the zero pair rule to simplify.
                             Then $-3+5=2$

1.   Use the chip model to add.

a. $-2+5$ d. $8+ \ -3$

b. $4+ \ -3$ e. $-6+ \ -3$

c. $-5+ \ -2$ f. $-6+4$

Subtraction of Integers using the two color chip model. Subtraction with integers follows the take-away model of subtraction. Always begin by modeling the minuend with chips, as demonstrated in the following examples. The subtrahend indicates the chips to be taken away. For example, let's use the chip model to subtract: $-3-2$.

 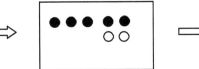 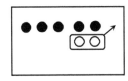

Represent the minuend -3 Introduce zero pairs to make   Take away 2.
                                 taking away 2 possible.   Then $-3-2=-5$

2. Use the chip model to subtract. The two color chips are on page A2.

    a. $-4-6$

    b. $3-7$

    c. $5--3$

    d. $-5--3$

    e. $-2--5$

3. A student uses the chip model to subtract. The results are shown in the diagram.

    What was the original subtraction problem?

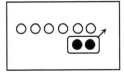

4. Use the chip model to find the answer to each problem. The two color chips are on page A2.

    a. $4-6$ and $4+-6$

    b. $-3-4$ and $-3+-4$

    c. $-5--2$ and $-5+2$

    d. $1--3$ and $1+3$

5. Do the following.

    a. What is the opposite of 6? What is the opposite of -3?

    b. What pencil-and-paper strategy do the results in Problem 4 suggest for subtracting integers?

    c. Illustrate your strategy with a few examples. Include a written explanation that convinces the reader that you know how to explain the strategy. Then let a student read your narrative and have them provide feedback on your work.

ACTIVITY 4.3.2   Multiplying and Dividing Integers

Material: sharp pencil   Name _____

This activity explores multiplying integers using the pattern model. It begins with known facts, and then uses patterns to derive new facts. Division with integers involves expressing division in terms of multiplication, and then building on known facts. The patterns lead to procedures for multiplication and division with integers that build on operations with whole numbers.

Multiplication of Integers: Use patterns of whole numbers and integers to determine the product 4×-3 :

A. $4 \times 3 = 12$ In the sequence of multiplications, there are two number patterns.

B. $4 \times 2 = 8$

     1. Describe one number pattern that you see.

C. $4 \times 1 = 4$

D. $4 \times 0 = 0$ **2.** Describe the other number pattern that you see.

E. $4 \times$ ____ = ____

F. $4 \times$ ____ = ____   **3.** Fill in the blanks of equation E using the information from the number patterns.

G. $4 \times$ ____ = ____

H. $4 \times$ ____ = ____   **4.** Complete equations F, G and H.

5. By the patterns above, $4 \times -3 =$ ____ .

6. Write a similar sequence of equations to discover the product 8×-4 .

7. Each equation $4 \times -3 = -12$, $3 \times -5 = -15$, and $7 \times -2 = -14$ can each be derived from the pattern model. Based on these equations, what is the sign of the product of a positive number and a negative number?

8. Do the following.

    a. If you know $4 \times 3 = 12$, how can you determine 4×-3 ?

    b. If you know $3 \times 5 = 15$, how can you determine 3×-5 ?

    c. Write a procedure for multiplying a positive integer and a negative integer.

9. Apply your procedure to find each product: 9×-2 and 6×-4 .

Use patterns of whole numbers and the integers to discover the product -5×-3 :

A. $-5 \times 3 = -15$ In the sequence of multiplications, there are two number patterns.

B. $-5 \times 2 = -10$

    10. Describe one number pattern that you see.

C. $-5 \times 1 = -5$

D. $-5 \times 0 = 0$ **11.** Describe the other number pattern that you see.

E. $-5 \times$ ____ = ____

F. $-5 \times$ ____ = ____   **12.** Fill in the blanks of equation E using the information from the number patterns.

G. $-5 \times$ ____ = ____

H. $-5 \times$ ____ = ____   **13.** Complete equations F, G and H.

14. By the patterns above, $-5 \times -3 = $ _____ .

15. Write a similar sequence of equations to discover the product -7×-4.

16. Each equation $-5 \times -3 = 15$, $-4 \times -3 = 12$, and $-7 \times -2 = 14$ can be derived from the pattern model. Based on these equations, what is the sign of the product of two negative numbers?

17. Do the following.

    a. If you know $4 \times 3 = 12$, how can you determine -4×-3?

    b. If you know $3 \times 5 = 15$, how can you determine -3×-5?

    c. Write a procedure for multiplying two negative integers.

18. Apply your procedure to find each product: -9×-2 and -6×-4.

Division of Integers: The definition of division for whole numbers naturally extends to integers: "Let a and b represent any integers with $b \neq 0$. Then $a \div b = c$ if and only if $a = b \times c$ for some unique integer c." The number c is known as the *quotient* or *missing factor*. The following steps illustrate how to apply the definition of division for integers to determine the quotient $120 \div -8$.

Step 1: We must find c such that $120 \div -8 = c$.

Step 2: Rewrite the division problem using the definition of division: $120 = -8 \times c$.

Step 3: According to the sign rules for multiplication with integers, c must be negative. The related whole number problem $120 = 8 \times \square$ is related to the whole number division problem $120 \div 8 = \square$. We know $120 \div 8 = 15$.

Step 4: Adjust the sign, as needed: $c = -15$.

Step 5: Restate the answer in terms of division: $120 \div -8 = -15$.

19. Use the definition of division to determine the quotients

    a. $-48 \div 12$ **b.** $-63 \div -7$ **c.** $72 \div -8$

20. Do the following.

    a. Use the equations $24 \div -3 = -8$, $35 \div -5 = -7$, and $80 \div -8 = -10$

        to write a *sign rule* that predicts the sign of the quotient when a positive integer is divided by a negative integer.

    b. Use the equations $-24 \div -3 = 8$, $-35 \div -5 = 7$, and $-80 \div -8 = 10$

        to write a *sign rule* that predicts the sign of the quotient when a negative integer is divided by a negative integer.

    c. Use the equations $-24 \div 3 = -8$, $-35 \div 5 = -7$, and $-80 \div 8 = -10$

        to write a *sign rule* that predicts the sign of the quotient when a negative integer is divided by a positive integer.

21. Apply the sign rules to divide.

    a. $-364 \div -26$ **b.** $-408 \div 24$ **c.** $1312 \div -41$

ACTIVITY 5.1.1 What is the Unit Fraction?

Material: sharp pencil and pattern blocks (p. A3)   Name _____

A fraction is a collection of equal-sized parts. For example, in the fraction $\frac{3}{7}$, the 3 indicates the fraction is a collection of three equal-sized parts and the 7 indicates that each equal-sized part is called one-seventh. In this case, the "whole unit" is comprised of seven equal-sized parts. A unit fraction is a fraction of the form $\frac{1}{n}$, such as $\frac{1}{2}$, $\frac{1}{3}$, or $\frac{1}{4}$. In this activity we illustrate how to use pattern blocks to give the student more experience with fractions. We also give you a chance to think about the meaning of fractions by presenting "reversal problems," where we model a fraction and ask you to draw the whole unit. Reversal problems require the ability to identify a unit fraction and apply the fact that a whole unit consists of *n* unit fractions of size $\frac{1}{n}$.

Refer to the pattern blocks on page A3 for questions 1-3. For questions 1-3, a whole unit is shown. Your goal is to write the listed shape as a unit fraction, if possible. For example, if 3 triangles fit into the whole without any gaps or overlaps, then the triangle is $\frac{1}{3}$.

1. The whole unit is shown.

    a. triangle

    b. trapezoid

    c. parallelogram

    whole unit for Probem 1

2. The whole unit is shown.

    a. triangle

    b. trapezoid

    c. parallelogram

    whole unit for Problem 2

3. The whole unit is shown.

    a. triangle

    b. trapezoid

    c. parallelogram

    whole unit for Problem 3

4. Identify the fraction represented by the dot in the rulers.

    a.   c.

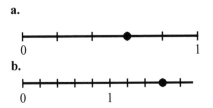

    b.   d.

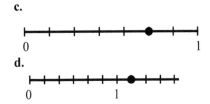

5.   The rectangle represents $\frac{2}{3}$.

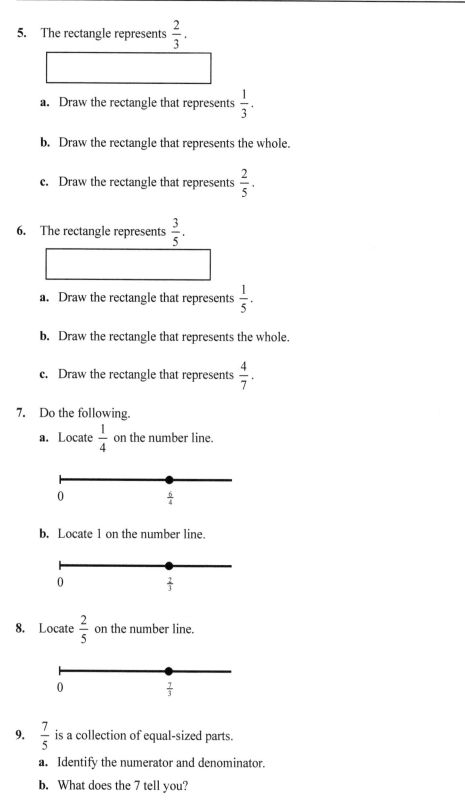

   a.   Draw the rectangle that represents $\frac{1}{3}$.

   b.   Draw the rectangle that represents the whole.

   c.   Draw the rectangle that represents $\frac{2}{5}$.

6.   The rectangle represents $\frac{3}{5}$.

   a.   Draw the rectangle that represents $\frac{1}{5}$.

   b.   Draw the rectangle that represents the whole.

   c.   Draw the rectangle that represents $\frac{4}{7}$.

7.   Do the following.

   a.   Locate $\frac{1}{4}$ on the number line.

     0                     $\frac{6}{4}$

   b.   Locate 1 on the number line.

     0                     $\frac{2}{3}$

8.   Locate $\frac{2}{5}$ on the number line.

     0                     $\frac{7}{3}$

9. $\frac{7}{5}$ is a collection of equal-sized parts.

   a.   Identify the numerator and denominator.

   b.   What does the 7 tell you?

   c.   What does the 5 tell you?

ACTIVITY 5.1.2 Unit Fractions and Serving Size

Material: sharp pencil   Name _____

The Nutrition Labeling and Education Act of 1990 regulates terms used on food packages. This act limits the type of health claims manufacturers can make about their products, and it requires manufacturers to provide formatted food labels for most prepared food products. The "reference amount customarily consumed" (RACC) for a prepared food is the amount of food that a person typically eats in one meal. It is determined by the Food and Drug Administration (FDA). For example, the reference amount customarily consumed for a heavy weight cake such as cheese cake is 125 grams, while the reference amount customarily consumed for a light weight cake such as angel food cake is 55 grams. The "serving size" is an allowable unit fraction of the product that most closely approximates the RACC. In this activity, we will learn how to calculate and represent the serving size of a product.

For food items that that are typically served in pieces, such as pizza, cake, or melon, manufacturers must state the serving size as an "allowable" unit fraction of the food product, choosing one of the unit fractions $\frac{1}{2}$, $\frac{1}{3}$, $\frac{1}{4}$, $\frac{1}{5}$, $\frac{1}{6}$, $\frac{1}{8}$, $\frac{1}{9}$, $\frac{1}{10}$, $\frac{1}{12}$, $\frac{1}{15}$, $\frac{1}{16}$, $\frac{1}{18}$, $\frac{1}{20}$, $\frac{1}{24}$, $\frac{1}{27}$, and $\frac{1}{30}$. The following table shows how the serving size would be determined for a heavy weight cake product weighing 538 grams. The FDA suggests that the RACC for heavy weight cake is 125 grams.

size of one slice of cake as a unit fraction	$\frac{1}{2}$	$\frac{1}{3}$	$\frac{1}{4}$
weight of one slice of cake that weighs 538 grams (g)	$538 \div 2 \approx 179$ g	$538 \div 3 \approx 135$ g	$538 \div 4 \approx 108$ g

135 g is closest to the suggestion of 125 g, so the serving size is stated as "$\frac{1}{3}$ (135 g)."

1. Calculate the serving size (which is an allowable unit fraction and the weight) for each product.
   a. The heavy weight cake weighs 1145 grams. The RACC for heavy food cake is 125 grams.
   b. The angel food cake weighs 320 grams. The RACC for angel food cake is 55 grams.
   c. The angel food cake weighs 910 grams. The RACC for angel food cake is 55 grams.
   d. The cheese cake weighs 1234 grams. The RACC for cheese cake is 125 grams.
   e. The cheese cake weighs 2600 grams. The RACC for cheese cake is 125 grams.

2. The reference amount customarily consumed for pizza is 140 grams.
   Calculate the serving size for a pizza with the given weight.
   a. 844 grams
   b. 1296 grams
   c. 2120 grams

3. How many servings are in an extra large pizza that weighs 3660 grams?

4. The nutrition label on a food product says that the serving size is $\frac{1}{5}$ (162 g). What is the weight of the food?

ACTIVITY 5.1.3   Draw a Diagram to Solve a Problem Involving Fractions

Material: sharp pencil   Name _____

The popular adage "A picture is worth a thousand words" applies to solving problems. A diagram helps the learner organize information. A diagram also evokes thought. This activity provides opportunities for you to solve problems with fractions using the *draw a diagram* strategy. The diagrams provide a way to tap into the meaning of fractions and solve problems without relying on mathematical procedures. Let's illustrate the strategy with an example:

Joseph has 24 red marbles. That is 3/5 of his marbles. How many marbles does Joseph have in all?

A fraction can be viewed a collection of equal-sized parts. For example, consider the fraction $\frac{3}{5}$. The 3 means that the fraction is a collection of 3 equal-sized parts. The 5 means that each equal-sized part is called $\frac{1}{5}$. So we draw a rectangle to represent the 24 marbles and split the rectangle into 3 equal-sized parts, as shown, where each part is $\frac{1}{5}$. Each equal-size part represents 8 marbles, because $24 \div 3 = 8$. Finally, one whole equals $\frac{5}{5}$.

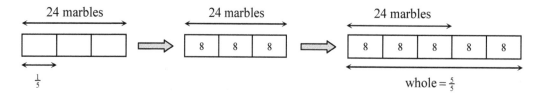

$5 \cdot 8 = 40$, so Joseph has 40 marbles in all.

Solve the following problems using the *draw a diagram* problem solving strategy.

1.   Sheila deleted $\frac{1}{3}$ of the songs on her playlist. She has 24 songs remaining on her playlist. How many songs does Sheila have on her playlist altogether?

2.   Saundra is a pitcher on a softball team. She struck out 168 batters during the season, which is $\frac{3}{7}$ of the batters that she faced. How many batters did she face during the season?

3.   The farmer loaded $\frac{2}{5}$ of the boxes onto the truck. He needed to load 345 more boxes of tomatoes onto the truck. How many boxes of tomatoes did he have to load altogether?

4.   Aaron spilled $\frac{3}{4}$ of the jelly beans in the bag. There were 336 jelly beans in the bag altogether. How many jelly beans did Aaron spill?

5.   Samantha ran $\frac{2}{5}$ of the race. She needs to run 240 more meters to finish the race. What is the length of the race?

6.   Three-eighths of the students in the club are boys. There are 24 more girls than boys in the club. How many students are there in the club altogether?

7.   A piggy bank contains pennies and nickels. Two-sevenths of the coins are pennies. There are 42 more nickels than pennies. How much money is in the piggy bank?

ACTIVITY 5.2.1   Adding and Subtracting Fractions with Fraction Strips

Material: sharp pencil, fraction strips (pp. A4-A7),   Name _____

and pattern blocks (p. A3)

Fraction strips are wonderful manipulatives to help students understand adding and subtracting fractions. They make it easier to see the need for a common denominator for adding and subtracting fractions. We begin with addition and subtraction with common denominators. Then we will help you answer the crucial question: how do students find a common denominator using fraction strips?

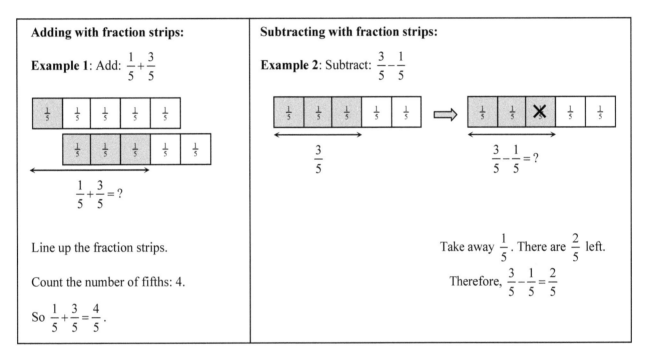

Adding with fraction strips:	Subtracting with fraction strips:
Example 1: Add: $\frac{1}{5}+\frac{3}{5}$	**Example 2**: Subtract: $\frac{3}{5}-\frac{1}{5}$
$\frac{1}{5}+\frac{3}{5}=?$	$\frac{3}{5}-\frac{1}{5}=?$
Line up the fraction strips.	Take away $\frac{1}{5}$. There are $\frac{2}{5}$ left.
Count the number of fifths: 4.	Therefore, $\frac{3}{5}-\frac{1}{5}=\frac{2}{5}$
So $\frac{1}{5}+\frac{3}{5}=\frac{4}{5}$.	

Add or subtract the following using the fraction strips. Draw pictures to show your work.

1. $\frac{3}{8}+\frac{5}{8}$ 4. $\frac{5}{8}-\frac{3}{8}$

2. $\frac{1}{6}+\frac{4}{6}$ 5. $\frac{3}{5}-\frac{1}{5}$

3. $\frac{3}{10}+\frac{4}{10}$ 6. $\frac{7}{12}-\frac{1}{12}$

After adding fractions that have a common denominator, the next step will be moving to fractions with unlike denominators. Students have to determine what the least common denominator will be. The fraction strips must be changed into equivalent strips that have overlapping cut lines for both strips.

7. Discover a method for students to find a common denominator between fourths and halves using the fraction strips.

8. Discover a method for students to find a common denominator between fourths and sixths using the fraction strips.

Example 3: Add: $\dfrac{1}{5}+\dfrac{1}{2}$

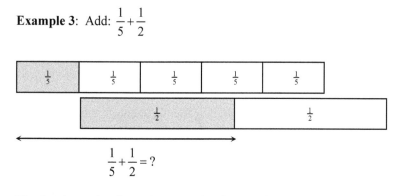

$$\dfrac{1}{5}+\dfrac{1}{2}=?$$

The fraction strips do not have overlapping cut lines, so we must change one or both to obtain the same cut lines so that it is possible to represent the sum as a fraction.

Then we see that $\dfrac{1}{5}+\dfrac{1}{2}=\dfrac{2}{10}+\dfrac{5}{10}$, which is a problem we already know how to solve: $\dfrac{1}{5}+\dfrac{1}{2}=\dfrac{2}{10}+\dfrac{5}{10}=\dfrac{7}{10}$.

Therefore, $\dfrac{1}{5}+\dfrac{1}{2}=\dfrac{7}{10}$.

Example 4: Subtract: $\dfrac{1}{2}-\dfrac{1}{5}$.

$$\dfrac{1}{2}-\dfrac{1}{5}=?$$

The fraction strips do not have the same cut lines. We must change one or both fraction strips so they have the same cut lines. In this case, both fractions are renamed: $\dfrac{1}{5}=\dfrac{2}{10}$ and $\dfrac{1}{2}=\dfrac{5}{10}$. Now we can begin with $\dfrac{5}{10}$ and then take away $\dfrac{2}{10}$.

$$\dfrac{1}{2}-\dfrac{1}{5}=\dfrac{5}{10}-\dfrac{2}{10}=\dfrac{3}{10}$$

Therefore, $\dfrac{1}{2}-\dfrac{1}{5}=\dfrac{3}{10}$.

Add or subtract the following using the fraction strips. Draw pictures to show your work.

9. $\dfrac{2}{3}+\dfrac{1}{6}$

10. $\dfrac{3}{4}+\dfrac{5}{6}$

11. $\dfrac{7}{10}-\dfrac{2}{5}$

12. $\dfrac{3}{4}-\dfrac{1}{3}$

13.   The hexagon represents 1. Write the unit fraction associated with each piece.

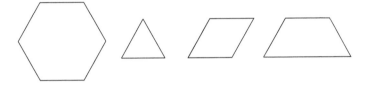

Use the hexagon to model and find each sum or difference.

14. $\dfrac{1}{2}+\dfrac{1}{6}$

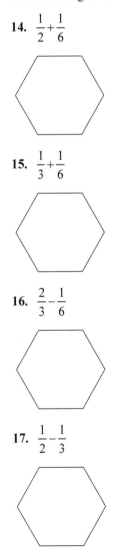

15. $\dfrac{1}{3}+\dfrac{1}{6}$

16. $\dfrac{2}{3}-\dfrac{1}{6}$

17. $\dfrac{1}{2}-\dfrac{1}{3}$

ACTIVITY 5.2.2   Solving Equations with Fractions: Part I

Material: sharp pencil   Name _____

This activity focuses on the reasons behind various steps of solving equations such as $n - \frac{3}{4} = \frac{7}{5}$ and $n + \frac{2}{3} = \frac{11}{6}$.

We demonstrate three ways to solve the equations using: (a) the definition of subtraction; (b) the fact that addition and subtraction are inverse operations; and (c) the basic properties of fractions. According to the definition of subtraction, the equations $\frac{a}{b} - \frac{c}{d} = \frac{e}{f}$ and $\frac{a}{b} = \frac{c}{d} + \frac{e}{f}$ are equivalent. This means the equations $n - \frac{3}{4} = \frac{7}{5}$ and $n = \frac{3}{4} + \frac{7}{5}$ are equivalent, and the equations $n + \frac{2}{3} = \frac{11}{6}$ and $n = \frac{11}{6} - \frac{2}{3}$ are equivalent.

Basic properties of addition and subtraction of fractions are: the commutative property of addition $\frac{a}{b} + \frac{c}{d} = \frac{c}{d} + \frac{a}{b}$; the associative property of addition $(\frac{a}{b} + \frac{c}{d}) + \frac{e}{f} = \frac{a}{b} + (\frac{c}{d} + \frac{e}{f})$; the additive identity property $0 + \frac{a}{b} = \frac{a}{b}$ and $\frac{a}{b} + 0 = \frac{a}{b}$; additive inverse property $\frac{a}{b} + \frac{-a}{b} = 0$ and $\frac{-a}{b} + \frac{a}{b} = 0$; adding the opposite property $\frac{a}{b} - \frac{c}{d} = \frac{a}{b} + \frac{-c}{d}$; and addition and subtraction are inverse operations so that $\frac{a}{b} + \frac{c}{d} - \frac{c}{d} = \frac{a}{b}$ and $\frac{a}{b} - \frac{c}{d} + \frac{c}{d} = \frac{a}{b}$. Of course, the properties of equality apply, too: if $\frac{a}{b} = \frac{c}{d}$, then $\frac{a}{b} + \frac{e}{f} = \frac{c}{d} + \frac{e}{f}$, by the addition property of equality; if $\frac{a}{b} = \frac{c}{d}$, then $\frac{a}{b} - \frac{e}{f} = \frac{c}{d} - \frac{e}{f}$, by the subtraction property of equality.

Problems 1-3 have been solved for you. Justify the steps by choosing a reason from the Word Bank:

- Commutative property of addition
- Associative property of addition
- Additive inverse property
- Addition property of equality
- Adding the opposite property
- Identity property of addition

- Definition of subtraction
- Rule for adding fractions
- Rule for subtracting fractions
- Simplification
- Subtraction property of equality
- Addition and subtraction are inverse operations

1. Solve $\frac{2}{3} + n = \frac{6}{7}$. Use the definition of subtraction.

Equation (E)	Reason (R)
E1. $\frac{2}{3} + n = \frac{6}{7}$	R1. Original equation
E2. $n = \frac{6}{7} - \frac{2}{3}$	R2. _____
E3. $n = \frac{6 \cdot 3 - 2 \cdot 7}{3 \cdot 7}$	R3. _____
E4. $n = \frac{4}{21}$	R4. _____

2. Solve $\dfrac{2}{3}+n=\dfrac{6}{7}$. Use the fact that addition and subtraction are inverse operations.

Equation	Reason
E1. $\dfrac{2}{3}+n=\dfrac{6}{7}$	R1. Original equation
E2. $\dfrac{2}{3}+n-\dfrac{2}{3}=\dfrac{6}{7}-\dfrac{2}{3}$	R2. _____
E3. $n=\dfrac{6}{7}-\dfrac{2}{3}$	R3. _____
E4. $n=\dfrac{6\cdot 3-2\cdot 7}{3\cdot 7}$	R4. _____
E5. $n=\dfrac{4}{21}$	R5. _____

3. Solve $\dfrac{2}{3}+n=\dfrac{6}{7}$. Use the basic properties of fractions.

Equation	Reason
E1. $\dfrac{2}{3}+n=\dfrac{6}{7}$	R1. Original equation
E2. $\dfrac{-2}{3}+\left(\dfrac{2}{3}+n\right)=\dfrac{-2}{3}+\dfrac{6}{7}$	R2. _____
E3. $\left(\dfrac{-2}{3}+\dfrac{2}{3}\right)+n=\dfrac{-2}{3}+\dfrac{6}{7}$	R3. _____
E4. $0+n=\dfrac{-2}{3}+\dfrac{6}{7}$	R4. _____
E5. $n=\dfrac{-2}{3}+\dfrac{6}{7}$	R5. _____
E6. $n=\dfrac{-2\cdot 7+3\cdot 6}{3\cdot 7}$	R6. _____
E7. $n=\dfrac{4}{21}$	R7. _____

4. Solve $n-\dfrac{4}{5}=\dfrac{7}{3}$.

   a. Use the basic properties of fractions.

   b. Use the definition of subtraction.

   c. Use the fact that addition and subtraction are inverse operations.

ACTIVITY 5.3.1   The Area Model for Multiplying Fractions

Material: sharp pencil   Name _____

In this activity, we will use the *area model* to find the product of two fractions. A *unit square* is a square that has length 1 unit and width 1 unit, where a "unit" could be, for example, an inch or centimeter. In the area model, the product $\frac{a}{b} \cdot \frac{c}{d}$ is the number of unit squares that fit in a rectangle having length $\frac{a}{b}$ units and width $\frac{c}{d}$ units, without gaps or overlaps. We will illustrate how to use the area model to express $\frac{6}{4} \cdot \frac{5}{3}$ as a fraction. Figure A shows a unit square and a rectangle with length $\frac{6}{4}$ units and width $\frac{5}{3}$ units.

Figure A. How many square units fit in the rectangle with length $\frac{6}{4}$ units and width $\frac{5}{3}$ units?

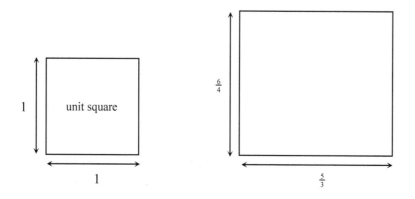

In Figure B, we partition the shapes using grid lines. The fractions in $\frac{6}{4} \cdot \frac{5}{3}$ suggest drawing grid lines of increments $\frac{1}{4}$ and $\frac{1}{3}$, as shown.

Figure B. Partition the shapes and count the number of equally-sized rectangles in the unit square and rectangle.

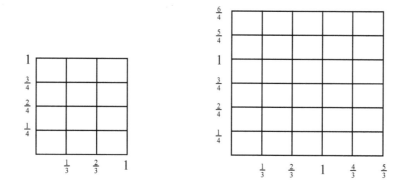

We see that $4 \cdot 3 = 12$ equal-sized rectangles fit in one unit square, and that $6 \cdot 5 = 30$ equal-sized rectangles fit in the array model of $\frac{6}{4} \cdot \frac{5}{3}$. The number of unit squares that fit in the array equals the number of groups of 12 in 30:

$30 \div 12 = \frac{30}{12}$. Then $\frac{6}{4} \cdot \frac{5}{3} = \frac{30}{12}$. Try this approach for $\frac{6}{5} \cdot \frac{7}{4}$. Turn the page to see the diagram that you should get.

Use the area model to determine $\frac{6}{5} \cdot \frac{7}{4}$.

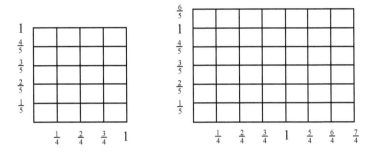

We see that $5 \cdot 4 = 20$ equal-sized rectangles fit in the unit square, and that $6 \cdot 7 = 42$ equal-sized rectangles fit in the rectangular array model of $\frac{6}{5} \cdot \frac{7}{4}$. The number of unit squares that fit in the rectangle equals the number of groups of 20 in 42: $42 \div 20 = \frac{42}{20}$. Then $\frac{6}{5} \cdot \frac{7}{4} = \frac{42}{20}$.

Use the area model to express each product as a fraction. Be sure to draw and partition the unit square and the model of the product of fractions. Do not simplify the answer.

1. $\dfrac{5}{4} \cdot \dfrac{3}{2}$

2. $\dfrac{4}{2} \cdot \dfrac{4}{3}$

3. $\dfrac{3}{4} \cdot \dfrac{5}{3}$

4. $\dfrac{2}{5} \cdot \dfrac{3}{4}$

5. Look for a pattern in the products to develop a rule for multiplying two fractions: $\dfrac{a}{b} \cdot \dfrac{c}{d}$.

ACTIVITY 5.3.2   Solving Equations with Fractions: Part II

Material: sharp pencil   Name _____

This activity focuses on the reasons behind various steps of solving equations such as $\frac{3}{4}n = \frac{7}{5}$ and $n \div \frac{2}{3} = \frac{11}{6}$. We demonstrate three ways to solve the equations using: (a) the definition of division; (b) the fact that multiplication and division are inverse operations; and (c) the basic properties of fractions. According to the definition of division, the equations $\frac{a}{b} \div \frac{c}{d} = \frac{e}{f}$ and $\frac{a}{b} = \frac{c}{d} \times \frac{e}{f}$ are equivalent. This means the equations $n \div \frac{2}{3} = \frac{11}{6}$ and $n = \frac{11}{6} \times \frac{2}{3}$ are equivalent, and the equations $\frac{3}{4}n = \frac{7}{5}$ and $n = \frac{7}{5} \div \frac{3}{4}$ are equivalent.

Basic properties of multiplication and division of fractions are: the commutative property of multiplication $\frac{a}{b} \times \frac{c}{d} = \frac{c}{d} \times \frac{a}{b}$; the associative property of multiplication $(\frac{a}{b} \times \frac{c}{d}) \times \frac{e}{f} = \frac{a}{b} \times (\frac{c}{d} \times \frac{e}{f})$; the identity property of multiplication $1 \times \frac{a}{b} = \frac{a}{b}$ and $\frac{a}{b} \times 1 = \frac{a}{b}$; the multiplicative inverse property $\frac{a}{b} \times \frac{b}{a} = 1$ and $\frac{b}{a} \times \frac{a}{b} = 1$; multiplication and division are inverse operations so that $\frac{a}{b} \times \frac{c}{d} \div \frac{c}{d} = \frac{a}{b}$ and $\frac{a}{b} \div \frac{c}{d} \times \frac{c}{d} = \frac{a}{b}$; and the invert and multiply rule $\frac{a}{b} \div \frac{c}{d} = \frac{a}{b} \times \frac{d}{c}$. Of course, the properties of equality apply, too: if $\frac{a}{b} = \frac{c}{d}$, then $\frac{a}{b} \times \frac{e}{f} = \frac{c}{d} \times \frac{e}{f}$, by the multiplication property of equality; if $\frac{a}{b} = \frac{c}{d}$, then $\frac{a}{b} \div \frac{e}{f} = \frac{c}{d} \div \frac{e}{f}$, by the division property of equality, provided $e \neq 0$.

Problems 1-3 have been solved for you. Justify the steps by choosing a reason from the Word Bank:

- Associative property of multiplication
- Commutative property of multiplication
- Definition of division
- Division property of equality
- Identity property of multiplication
- Multiplication and division are inverse operations

- Multiplicative inverse property
- Multiplication property of equality
- Rule for multiplying fractions
- Simplification
- The invert and multiply rule

1. Solve $\frac{3}{5}n = \frac{6}{7}$. Use the definition of division.

Equation (E)	Reason (R)
E1. $\frac{3}{5}n = \frac{6}{7}$	R1. Original equation
E2. $n = \frac{6}{7} \div \frac{3}{5}$	R2. _____
E3. $n = \frac{6}{7} \times \frac{5}{3}$	R3. _____
E4. $n = \frac{6 \times 5}{7 \times 3}$	R4. _____
E5. $n = \frac{30}{21}$	R5. _____
E6. $n = 1\frac{3}{7}$	R6. _____

2. Solve $\dfrac{3}{5}n = \dfrac{6}{7}$. Use the fact that multiplication and division are inverse operations.

Equation (E)	Reason (R)
E1. $\dfrac{3}{5}n = \dfrac{6}{7}$	R1. Original equation
E2. $\dfrac{3}{5}n \div \dfrac{3}{5} = \dfrac{6}{7} \div \dfrac{3}{5}$	R2. _____
E3. $n = \dfrac{6}{7} \div \dfrac{3}{5}$	R3. _____
E4. $n = \dfrac{6}{7} \times \dfrac{5}{3}$	R4. _____
E5. $n = \dfrac{6 \times 5}{7 \times 3}$	R5. _____
E6. $n = \dfrac{30}{21}$	R6. _____
E7. $n = 1\dfrac{3}{7}$	R7. _____

3. Solve $\dfrac{3}{5}n = \dfrac{6}{7}$. Use the basic properties of fractions.

Equation (E)	Reason (R)
E1. $\dfrac{3}{5}n = \dfrac{6}{7}$	R1. Original equation
E2. $\dfrac{5}{3} \times \left(\dfrac{3}{5}n \right) = \dfrac{5}{3} \times \dfrac{6}{7}$	R2. _____
E3. $\left(\dfrac{5}{3} \times \dfrac{3}{5} \right)n = \dfrac{5}{3} \times \dfrac{6}{7}$	R3. _____
E4. $1 \times n = \dfrac{5}{3} \times \dfrac{6}{7}$	R4. _____
E5. $n = \dfrac{5}{3} \times \dfrac{6}{7}$	R5. _____
E6. $n = \dfrac{5 \times 6}{3 \times 7}$	R6. _____
E7. $n = \dfrac{30}{21}$	R7. _____
E8. $n = 1\dfrac{3}{7}$	R8. _____

4. Solve $n \div \dfrac{2}{5} = \dfrac{7}{3}$.

    a. Use the basic properties of fractions.

    b. Use the definition of division.

    c. Use the fact that multiplication and division are inverse operations.

ACTIVITY 5.4.1 Diagrams and Ratios

Material: sharp pencil and straightedge   Name _____

A ratio is a way to compare two quantities. There are three ways to represent the ratio "three to five" symbolically: 3:5, 3 to 5, and $\frac{3}{5}$. In this activity, we focus on ways to represent ratios and their relationships with a diagram. For example, suppose Maria has 3 coins for every 5 coins that Krista has. The following diagram provides a visualization of the relationship.

M ☐ ☐ ☐
K ☐ ☐ ☐ ☐ ☐

We see that there are three of these to five of that. The diagram does not tell exactly how many coins they each have because a ratio is a comparison. The diagram can be modified with additional information.

Maria has 3 coins for every 5 coins that Krista has. Krista has 20 more coins than Maria.	Maria has 3 coins for every 5 coins that Krista has. Altogether, they have 96 coins.
M ☐ ☐ ☐ K ☐ ☐ ☐ ☐ ☐ ←——→ 20	M ☐ ☐ ☐ K ☐ ☐ ☐ ☐ ☐ ↕ 96

1. Write two ratios for the diagram.

2. Two local politicians, Culligan and Mulligan, competed for a seat on the city council. Culligan spent $6420 more than Mulligan in advertising during the campaign. Culligan spent $7 for every $4 dollars Mulligan spent. How much money did each politician spend? Solve this problem using a diagram.

3. Samantha read three pages for every seven pages Annie read. Samantha read 84 fewer pages than Annie. How many pages did Samantha read? Solve this problem using a diagram.

4. There were 3 cats for every 5 mice in the barn. Altogether, there were 112 cats and mice in the barn. How many cats were in the barn? Solve this problem using a diagram.

5. Marco has five coins for every seven coins Ted has. Altogether, they have 492 coins. How many coins do they each have? Solve this problem using a diagram.

6. If Maria had 4 more coins, then Krista would have five coins for every three coins Maria had. Altogether, they had 140 coins. How many coins do they each have? Solve this problem using a diagram.

7. Graph each pair of ratios. For each ratio *a:b*, plot the first term *a* along the horizontal axis and the second term *b* along the vertical axis. The use a straightedge to draw a line through the two points that represent the two ratios.

    a. 2:3 and 4:6 (which are equivalent ratios)

    b. 3:5 and 6:10 (which are equivalent ratios)

    c. 2:5 and 3:8 (which are not equivalent ratios)

    d. 4:7 and 5:8 (which are not equivalent ratios)

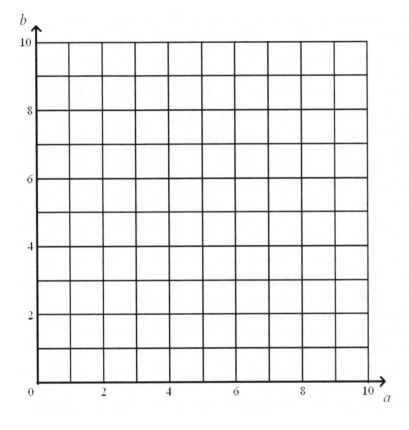

8. What did you notice about the equivalent ratios in the graph in Problem 7?

ACTIVITY 5.4.2 Unit Prices and the Better Buy

Material: sharp pencil   Name _____

Shoppers seem more conscious of how they spend their money during difficult economic times. Sales are frequented and pennies are pinched. Shoppers can often save money by purchasing items in large quantities. This activity explores shopping and cost per unit.

The newspaper advertisement shown says "Carpet Special .99¢ per square foot."
How much carpet can you buy for $1 without including tax?

If you said one square foot, then please review the price and state precisely the price per square foot. According to the advertisement, you can buy 101 square feet of carpet for $1 because the price is less than one cent per square foot and $100 \div 0.99 \approx 101.01$. What a bargain, the carpet is nearly free! The advertisement should say $0.99 per square foot or 99¢ per square foot. If you are charging cents per item, then make sure the decimal point is in the correct place.

You need laundry detergent for your high efficiency washer.
Compare the following prices of Brand A in price per ounce.

Brand A	Size	Cost	Price per ounce
Large	120 ounces	$10.99	1.
Medium	90 ounces	$8.99	2.
Small	55 ounces	$5.99	3.

4. Which is the better buy, based on the price per ounce?

While reading the labels, you notice the smaller bottle is actually triple concentrated. Each bottle states the number of loads the bottle will wash. So you decide to compare the detergents by the number of loads they can wash. Compare the following in price per load.

Brand A	Loads	Cost	Price per load
Large Bottle	120 loads	$10.99	5.
Medium Bottle	85 loads	$8.99	6.
Small Bottle	100 loads	$5.99	7.

8. What is the better buy, based on the price per load?

As you reach the aisle with the dry cereal, you decide to compare three brands A, B, and C of oat cereal. Calculate the price per ounce for each brand.

Name	Size	Cost	Price per ounce
Brand A	21.4 ounces	$3.29	9.
Brand B	36.1 ounces	$4.99	10.
Brand C	18.4 ounces	$3.09	11.

12. What cereal is the better buy?

13. Brand A of oat cereal has a coupon for $1 off the total price if you buy two boxes.
    If you use the coupon, which cereal will be the better buy?

Then you check the eggs in the refrigerator section. Compare the following.

Name	Cost	Price per egg
Dozen Eggs	$1.19	14.
18 count eggs	$1.59	15.
5 dozen eggs	$3.19	16.

17. What is the better buy for eggs?

18. If your family had the choice, which would you buy?

19. Discuss a situation where you will typically prefer to buy the store brand than the corresponding name brand.

20. Discuss a situation where you will typically prefer to buy the name brand than the corresponding store brand.

21. On the way home from the grocery store you pass a sign from the pizza shop that says, "Two medium pizzas for $8 or one large pizza for $10." Using the area formula for a circle $A = \pi r^2$, compare the cost of the pizzas per square inch.

Size	Diameter	Cost	Cost per square inch
Large	20 inches	$10 for 1	22.
Medium	10 inches	$8 for 2	23.

24. Which is the better buy for pizza?

ACTIVITY 6.1.1   Modeling, Reading, and Writing Decimal Numbers

Material: sharp pencil   Name _____

In this activity, we will model decimals with 10-by-10 arrays and practice representing decimal numbers.

1. The diagrams model the place values 1 (*one*), 0.1 (read as *one tenth*) and 0.01 (read as *one hundredth*).

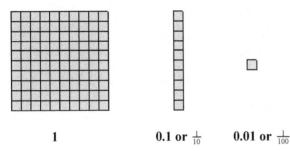

  1   **0.1 or** $\frac{1}{10}$ **0.01 or** $\frac{1}{100}$

Write the number represented by the shaded region in its fraction form and decimal number form.

a.   b.   c.

2. There are several ways to represent a whole number, for example, 345 = 3 hundreds, 4 tens, 5 ones, or 345 = 34 tens, 5 ones, or 345 = 345 ones. The multiple representations were important for regrouping needed for operations with whole numbers. There are several ways to represent decimal numbers, too. Use the diagrams to help you fill in the blanks.

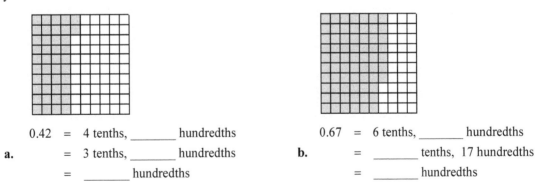

a.
  0.42 = 4 tenths, _____ hundredths
       = 3 tenths, _____ hundredths
       = _____ hundredths

b.
  0.67 = 6 tenths, _____ hundredths
       = _____ tenths, 17 hundredths
       = _____ hundredths

3. Model each decimal number using appropriate shading.

  a. 0.24   b. 0.72   c. 0.45

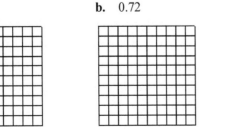

4. 63,459.1278 has four decimal digits, while 45.002 has three decimal digits. Determine the number of decimal digits.

    a. 0.45   **b.** 0.038   **c.** 0.00821   **d.** 521.33

5. The teacher asked her students to determine the number of decimal digits in the number 92.038. Mike said that it has three decimal digits and Neal said that it has five decimal digits. Mike gave the correct answer. Explain how each student arrived at their answer. What advice would you give Neal?

6. Each decimal number has two parts: the integer part and the decimal part. For example, 63,459.1278 has integer part 63,459 and decimal part 1278. The decimal number 63,459.1278 has the *short word form* "63 thousand, 459 and 1,278 ten-thousandths" and the *word form* "sixty-three thousand, four hundred fifty-nine and one thousand, two hundred seventy-eight ten-thousandths." The key to naming the fractional part is to determine the place value of the rightmost digit of the decimal part. For each decimal number, write the short word form and the word form.

    a. 6.34   **b.** 27.034   **c.** 7215.0034

7. Discuss a strategy for reading decimal numbers such as 0.45, 0.038, and 0.00821.

8. Discuss a strategy for reading decimal numbers such as 7.45, 92.038, and 4.00821.

9. Complete the table.

word form	short word form	standard form
a.		4.67
b.	72 thousand, 148 and 15 thousandths	
c. three thousand, twenty-eight millionths		
d. two thousand, sixteen and thirty five ten-thousandths		
e. two hundred thousand and forty-two thousandths		

10. Represent each decimal number as a mixed number.

    a. 42.638   **b.** 35.0027   **c.** 528.0405

11. There are several expanded forms of a decimal number. Complete the table.

standard form	4.237		8.604
expanded form	$4 + 0.2 + 0.03 + 0.007$		$8 + 0.6 + 0 + 0.004$
expanded form with fractions	$4 + \dfrac{2}{10} + \dfrac{3}{100} + \dfrac{7}{1000}$	$90 + 5 + \dfrac{1}{10} + \dfrac{8}{100} + \dfrac{2}{1000}$	
expanded form with multiplication	$4 \cdot 1 + 2 \cdot 0.1 + 3 \cdot 0.01 + 7 \cdot 0.001$		
expanded form with exponents	$4 \cdot 10^0 + 2 \cdot 10^{-1} + 3 \cdot 10^{-2} + 7 \cdot 10^{-3}$		

ACTIVITY 6.1.2   The Area Model for Multiplying Two Decimals

Material: sharp pencil   Name _____

In this activity, we will use the *area model* to find the product of two decimal numbers. Figure 1 shows a *square unit*, which is a square that has length 1 unit and width 1 unit, where a "unit" could be, for example, an inch or centimeter. Each small square is one hundredth of a unit square, and every 100 hundredths equals 1 unit square. Figure 2 shows a shaded rectangle with length 0.6 units and width 0.8 units. It is an array with 6 rows and 8 columns. The product $0.6 \cdot 0.8$ is the number of unit squares (see Figure 1) that fit in a rectangle with length 0.6 units and width 0.8 units (see Figure 2) without gaps or overlaps. We see that 48 hundredths fit in a rectangle with length 0.6 units and width 0.8 units. In other words, 0.48 unit squares fit in the rectangle. Then $0.6 \cdot 0.8 = 0.48$.

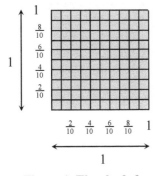

Figure 1. The shaded region is a model of 1 unit square.

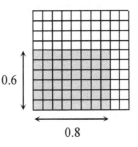

Figure 2. The shaded region is a model of $0.6 \cdot 0.8$

1. Use a diagram to model and determine each product.

    a. $0.3 \cdot 0.4$ **b.** $0.4 \cdot 0.7$ **c.** $0.6 \cdot 0.5$

The area model can also be used to model products of decimals such as $1.3 \cdot 2.4$. We interpret the product $1.3 \cdot 2.4$ as the number of unit squares (see Figure 1) that fit in the rectangle with length 1.3 units and width 2.4 units. The shaded rectangle is an array with 13 rows and 24 columns, and $13 \cdot 24 = 312$. So 312 hundredths fit in a rectangle with length 1.3 units and width 2.4 units. Note that 312 hundredths equals 3.12. Then $1.3 \cdot 2.4 = 3.12$.

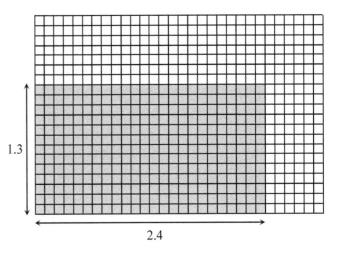

2. Use the area model to find each product.

Remember: Each tiny square is one hundredth and every 100 hundredths equal 1.

a. 1.4 • 1.6

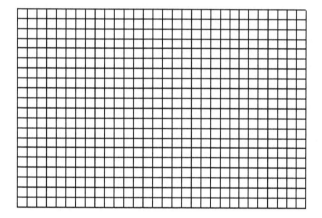

b. 1.5 • 2.3

c. 1.2 • 2.7

ACTIVITY 6.1.3 Reinventing the Standard Algorithm for Multiplying Decimals

Material: sharp pencil   Name _____

This activity sheds light on the underlying steps for the standard algorithm (procedure) for multiplying decimals. To build on what we know, we will convert the decimal numbers to fractions, and then multiply the fractions. Of course, when we divide by divide by positive powers of 10 (such as 10, 100, and 1000), we simply shift the decimal point. Then we will analyze and retrace our steps to "reinvent" an efficient algorithm for multiplying two decimals.

Multiply 1.3 and 0.04.

Step 1: Convert each decimal to its fraction form: $1.3 \cdot 0.04 = \dfrac{13}{10} \cdot \dfrac{4}{100}$.

Step 2: Multiply the fractions: $\dfrac{13}{10} \cdot \dfrac{4}{100} = \dfrac{13 \cdot 4}{10 \cdot 100}$.

Step 3: Simplify the numerator and denominator: $\dfrac{13 \cdot 4}{10 \cdot 100} = \dfrac{52}{1000}$.

Step 4: Convert the fraction form of the answer to a decimal number: $\dfrac{52}{1000} = 0.052$.

Let's retrace and analyze our steps.
- In **Step 2**, we are multiplying 13 and 4, and it seems like we temporarily ignored the decimal points in 1.3 and 0.04.
- In **Step 3**, we are dividing the product 52 by 1000. Note that 1000 has three zeros, and the sum of the number of decimal digits in 1.3 and 0.04 is three ($1 + 2 = 3$).
- In **Step 4**, we know that division by a positive power of 10 (such as 10, 100, or 1000) involves shifting the decimal point to the left. In particular, dividing by 1000 requires shifting the decimal point three places (or place values) to the left: $\underset{3 \; 2 \; 1}{\underbrace{.052.}}$ Then $\frac{52}{1000}$ equals 0.052.

Multiply the two decimal numbers as outlined in Steps 1-4 for $1.3 \cdot 0.04$

1. $2.05 \cdot 0.3$

2. $0.17 \cdot 0.105$

3. $15.2 \cdot 0.0016$

4. Write an efficient procedure for multiplying two decimal numbers. Use the example $4.25 \cdot 6.703$ to illustrate your procedure. You may want to incorporate the example in your writing. Ask another student to read your work and provide written comments on your paper.

ACTIVITY 6.2.1   Repeating Decimals

Material: sharp pencil   Name _____

A **rational number** is a number that can be expressed in the form $\frac{a}{b}$, where a and b are integers and $b \neq 0$. Some

fractions can be expressed as terminating decimal numbers, such as $\frac{314}{100} = 3.14$ or $68.027 = \frac{680,027}{1000}$. Some fractions can be

expressed as repeating decimal numbers, such as $\frac{5}{6} = 0.83333...$ or $\frac{35}{27} = 1.296296....$ In this activity, you will learn how

to categorize a rational number as a terminating decimal number or repeating decimal number. You will also learn how

to represent a repeating decimal number as a fraction.

1.   Give two examples of terminating decimals. Then express each one as a fraction. Then simplify the fraction.

2.   What is a terminating decimal?

3.   Examine the following fractions. Find their decimal equivalents.

   a. $\frac{1}{4}$

   b. $\frac{4}{10}$

   c. $\frac{3}{8}$

   d. $\frac{6}{75}$

   e. $\frac{6}{150}$

4.   Examine the following fractions. Simplify each fraction, and then find the prime factorization of the denominator of

     each simplified fraction.

   a. $\frac{1}{4}$

   b. $\frac{4}{10}$

   c. $\frac{3}{8}$

   d. $\frac{6}{75}$

   e. $\frac{6}{150}$

5.   The numbers 8, 25, and 50 have only 2 or 5 as prime factors, because $8 = 2^3$, $25 = 5^2$, and $50 = 2 \cdot 5^2$.

     Write two other numbers that have only 2 or 5 as prime factors.

6.   Suppose the common factor of a and b is 1 and $b > 1$. According to Theorem 1 in the textbook, the rational number

     $\frac{a}{b}$ will be equivalent to a terminating decimal number as long as only two specific prime numbers are possibly

     factors of b. What are those two specific prime numbers?

7.   Decide if the fraction is a terminating or non-terminating fraction without using a calculator.

   a. $\dfrac{3 \cdot 7}{2^3 \cdot 5}$ c. $\dfrac{63}{2 \cdot 2 \cdot 3 \cdot 3 \cdot 7 \cdot 5}$ e. $\dfrac{6}{14}$ g. $\dfrac{15}{24}$

   b. $\dfrac{4 \cdot 7}{2^3 \cdot 5 \cdot 7 \cdot 7}$ d. $\dfrac{12}{15}$ f. $\dfrac{7}{40}$ h. $\dfrac{a}{20}$, where a some counting number.

8. A *repeating decimal* is a non-terminating decimal with a repeating group of digits, such as 5.68747474... . The *repetend* of 5.68747474... is 74, and the *period* of 5.68747474... is 2. Rather than use the ellipsis (...) to denote the pattern continues, we can use a bar, and write 5.68747474... as $5.68\overline{74}$.

    a. Determine the repetend and period of the number 25.9903670367... .

    b. Express 25.9903670367... without using the ellipsis.

    c. Determine the repetend and period of the number $2.5\overline{807}$.

Now we demonstrate possible steps to represent the repeating decimal 2.58787... as a fraction.

- First, create the equation $n = 2.58787...$.

- Second, multiply n by an appropriate power of 10 so that the product has the repetend 87. Then $10n = 25.8787...$.

- Third, multiply n by another power of 10 so that the product has the repetend 87. Then $1000n = 2587.8787...$.

- Then both products $10n$ and $1000n$ have the same repetend, so we subtract to eliminate the fraction part:

$$
\begin{aligned}
1000n - 10n &= 2587.8787... - 25.8787... \\
990n &= 2587 - 25 \\
990n &= 2562 \\
n &= \frac{2562}{990} \\
2587.8787... &= \frac{2562}{990} \qquad \text{(Note: there is no need to simplify the fraction.)}
\end{aligned}
$$

9. Express the repeating decimals as a fraction.

    a. $0.\overline{4}$

    b. 0.121212...

    c. $0.2\overline{75}$

    d. $4.1\overline{2}$

    e. 2.627575...

    f. 56.046234234...

10. Do the following.

    a. Use a calculator to determine the decimal form of $\frac{185}{990}$.

    b. Use a calculator to determine the decimal form of $\frac{285}{990}$.

    c. Use a calculator to determine the decimal form of $\frac{385}{990}$.

    d. Use the results in parts (a)-(c) to predict the decimal form of $\frac{485}{990}$.

    e. Predict the decimal form of $\frac{585}{990}$.

    f. Use a calculator to determine the decimal form of $\frac{585}{990}$.

ACTIVITY 6.2.2   Irrational Numbers

Material: sharp pencil   Name _____

Terminating decimals and repeating decimals can be written as fractions. For example, $6.23 = \frac{623}{100}$ and $0.08\overline{5}... = \frac{85}{990}$.
Can all decimal numbers be written as fractions? Can you think of a decimal number that does not terminate or repeat?
What category does this decimal number belong to? This activity will discuss another category of decimal numbers
besides terminating and repeating, called irrational numbers.

Irrational numbers are decimal numbers that neither terminate nor repeat. One way is to create an irrational number is
to create a decimal number with infinitely decimal digits that has a pattern but does not have a repetend. For example:
3.45050050005… is an irrational number.

1.   Can you predict the next five digits for 3.45050050005… ?

2.   Can you predict the next five digits for the irrational number 0.171171117… ?

3.   Do the following.

   a.   Predict the next five digits of the decimal number 74.626226222… .

   b.   Explain why you think 74.626226222… is an irrational number.

4.   Create an irrational number that has a pattern but does not have a repetend.

5.   The numbers 1, 4, 9, and 25 are perfect squares. List three more perfect squares.

Another way to create an irrational number is to take the square root of a non-perfect square number. For example, the
square root of 2, written $\sqrt{2}$, is an irrational number. Using a calculator, we can round $\sqrt{2}$ to the nearest ten-thousandth
to get $\sqrt{2} \approx 1.4142$.

6.   Classify the number as a rational number or irrational number. If the number is a rational number, then write it in
     decimal form. If the number is an irrational number, then write it in decimal form to the nearest ten-thousandth.

   a. $\sqrt{7}$

   b. $\frac{6}{15}$

   c. $\sqrt{120}$

   d. $\sqrt{81}$

   e. $\frac{17}{66}$

7.   Find a counting number n such that $\sqrt{n}$ is an irrational number. Tell how you know $\sqrt{n}$ is an irrational number.

ACTIVITY 6.3.1   Using a Bar Diagram to Solve Percent Problems

Material: sharp pencil   Name _____

Percent is a multifaceted concept with strong connections to ratios, fractions, proportions, variables, representation, and equations. 35% and 150% are examples of percents. What is a percent? We can view a *percent* as a symbolic representation of a fraction in which the denominator is 100 (for example, $32\% = \frac{32}{100}$). Percents help us make informed decisions in a world that uses percents in polls, news reports, commission, interest, taxes, gratuities, probability, markups, discounts, and grades. In this activity, we use a bar diagram to set up a proportion to solve percent problems.

Here's an example: There are 240 sixth graders in the school district. 30% of them ride their bikes to school. How many sixth graders ride their bikes to school?

number of 6th graders

| | n | | 240 |

percent   30   100

Then $\frac{n}{30} = \frac{240}{100}$. Then $100n = 30 \cdot 240$. Then $100n = 7200$. Then $n = 7200/100 = 72$. So 72 sixth-graders ride their bikes to school.

The bar diagram

- provides a visual representation of the information
- helps students analyze the underlying structure of contextualized problems
- promotes reasoning in percent problems
- allows students to grasp the percent problem.

A bar diagram has the following form:

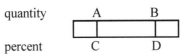

quantity   A   B

percent   C   D

C or D will be 100, depending on the problem, and three of the letters will be assigned numbers so that there is just one variable in the diagram. The 100% refers to the "whole" or basis for comparison. The diagram leads to one of the following proportions: part-to-part $\frac{A}{C} = \frac{B}{D}$ or part-to-whole $\frac{A}{B} = \frac{C}{D}$; either proportion leads to the same answer.

1. Figure 1 shows benchmark percents of $16. The benchmark percents allow students to use intuition, patterns, and experience with partitioning to create a bar diagram. Students should know 25%, 50%, and 75% of a number. Use patterns to write the missing numbers in the diagram.

Figure 1. Use a diagram to visualize a percent of $16

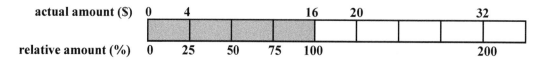

Any fractions formed by corresponding parts leads to a proportion. For example, $\frac{4}{25} = \frac{16}{100}$ because $4 \cdot 100 = 25 \cdot 16$. To use the bar diagram successfully, we need to make sure the numbers in each row of the bar diagram are in increasing order.

For each problem, draw the diagram, set up a proportion based on the diagram, and then solve the problem.

2. Paul purchased 650 plants, and 273 of them were tomato plants. What percent of the plants were tomato plants?

3. A teacher surveyed 275 students. 44% of the students said they preferred to dip their cookies in milk. How many students in the survey preferred to dip their cookies in milk?

4. Ellie bought a ticket to the zoo for $52. She paid 80% of the regular price. What is the regular price for a ticket to the zoo?

5. A newspaper article reported that 2015 people went to the county fair this year. That is 130% of the number of people who went to the fair last year. How many people went to the fair last year?

6. A farmer picked 450 apples. He noticed that 171 of the apples that he picked were bruised from the recent storm. What percent of the apples that he picked were bruised?

7. Wally has a collection of beads. 64% of the beads are silver. She has 336 silver beads. How many beads are in her collection?

8. 432 people said they saw the play this week. That is 120% of the number of people who saw the play last week. How many people saw the play last week?

9. 385 chickens on the farm clucked every morning at sunrise. That was 28% of all the chickens on the farm. How many chickens were on the farm?

10. 680 people said they voted in the election. 238 people said they voted for the candidate who promised free calculator batteries for students. How percent of the people voted for the candidate?

Activity 7.1.1 Multiple Representations of Functions

Material: sharp pencil   Name _____

We can represent a function using a table, a graph, an equation, or verbal rule. The concept of function, which we use to relate two variables, is a unifying idea because it integrates many mathematical topics: tables, graphs, variables, algebraic expressions, representations, quantifying relationships with equations, arithmetic operations and their properties, evaluating expressions, and solving equations. Functions and graphs are crucially important because they afford students powerful opportunities to comprehend advanced mathematics. The equation $y(x) = 3x - 2$ uses function notation and tells the reader how the variable y depends on the variable x, where x is the independent variable (input) and y is the dependent variable (output). This activity will be exploring each of these different representations by converting a function from one representation to another.

1. Use the following verbal rule to fill in the following table of values.
    "To find the output, multiply the input by five, then add two."

x	−2	−1	0	1	2
y					

2. "The length is four feet less than three times the width" is a verbal rule for a function.
    Determine the equation that relates the length l and the width w of the rectangle. Use function notation.

3. Describe in words how the variables x and y are related.
    Then graph the ordered pairs on a rectangular coordinate system.

x	−1	0	1	2	3	4
y	−5	−3	−1	1	3	5

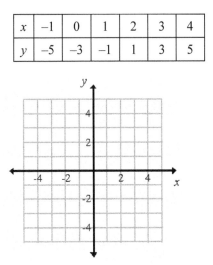

4. The graph shows a set of ordered pairs that determine a function. Represent the ordered pairs with a table.

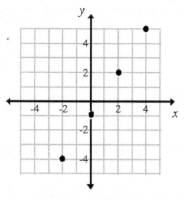

5. The following table is based on an equation. Complete the table, and then determine an equation that describes how the variables x and y are related.

x	0	1	2	3	4	5
y	0	1	4	9		

6. Restate the following linear function using a verbal rule: $y = 2x - 1$.

7. A homeowner obtains an estimate of the cost to build a fence. The cost C is given by $C(h) = 600 + 40h$ dollars, where h is the number of hours it takes to build the fence.
    a. What is the meaning of the 600 in the equation?
    b. What is the cost for each hour of labor to build the fence? Justify your answer.
    c. What does the ordered pair (4, 760) mean?
    d. The homeowner paid $960. How many hours did it take to build the fence?

8. There are 5 cars for every 8 trucks in the parking lot. Let c represent the number of cars and let t represent the number of trucks in the parking lot.
    a. Write an equation that relates the variables using function notation.
    b. Tell how you know your equation is correct.

ACTIVITY 7.1.2   Function Notation

Material: sharp pencil   Name _____

Functions and graphs are crucially important because they afford students powerful opportunities to comprehend advanced mathematics. This activity provides you opportunities to relate two variables using function notation.

1.   Use function notation to represent each relationship.

   a. Mike has 5 more marbles than Paul.

   b. The length of the rectangle is 3 cm fewer than 4 times the width of the rectangle.

2.   A homeowner uses light and dark bricks to make a patio with n rows and $n - 2$ columns.
     The dark bricks form a border around the light bricks, as shown.

     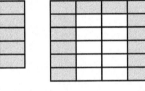

    5 rows, 3 columns   6 rows, 4 columns   7 rows, 5 columns

   a. What does the expression $3n - 4$ tell you?

   b. Use function notation and write an equation that expresses the number of dark bricks d in a n-by-$(n - 2)$ patio as a function of n.

   c. How many dark bricks are in a 40-by-38 patio?

   d. A patio has 674 dark bricks. What are the dimensions of the patio?

   e. Is it possible to build a patio with exactly 500 dark bricks? Justify your answer.

   f. How many light bricks are in a 28-by-26 patio?

   g. A patio has 221 black bricks. How many light bricks does the patio have?

3.   A taxi charges $2.80 for the first fifth of a mile, and then $0.35 each successive fifth or part of a mile.

   a. Verify that the cost for a 1-mile cab ride is $4.20.

   b. Verify that the cost for a 2-mile cab ride is $5.95.

   c. What is the cab fare for a 3-mile ride?

   d. Use function notation and express the cab fare C as a function of the number of miles m to complete the trip.

   e. Verify that $C(1)$ agrees with your answer in part (a).

   f. Verify that $C(2)$ agrees with your answer in part (b).

   g. Use the function to calculate the cab fare for a 15-mile taxi ride.

   h. Suppose the cab fare was $39.20. How many miles was the taxi ride?

ACTIVITY 7.1.3   Garden Design

Material: sharp pencil   Name _____

Jennifer wants to make two square gardens with walkways around and between them as depicted in the three designs. The walkways will be constructed with 1 ft-by-1 ft stepping stones.

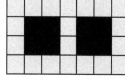

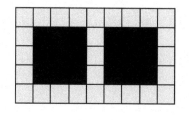

| width = 5 ft | width = 7 ft | width = 9 ft |
| length = 3 ft | length = 4 ft | length = 5 ft |

1. If the width of the design is 21 feet, what is the length of the design?

2. If the width of the design is 21 feet, what is the area of design?

3. If the width of the design is 21 feet, what is the length and width of each square garden?

4. If the width of the design is 21 feet, what is the total area of the square gardens?

5. If the width of the design is w feet, what is the length of the design?

6. If the width of the design is w feet, what is the area of the design?

7. If the width of the design is w feet, what is the length and width of each garden?

8. Determine the total area g of the two gardens as a function of the width w of the design.

9. If the width of the design is 17 feet, what is the total area of the gardens?

10. If the area of the two gardens is 128 square feet, what is the width of the design?

11. Is it possible to have the area of the two gardens exactly 325 square feet? Explain.

12. Is it possible to have the area of the two gardens exactly 300 square feet? Explain.

13. Is it possible to have the area of the two gardens exactly 800 square feet? Explain.

ACTIVITY 7.2.1   Three Methods to Solve a System of Two Linear Equations

Material: sharp pencil   Name _____

When two linear equations with the same two variables are presented together, such as $x + 4y = 10$ and $-3x + 7y = 2$, they are called a **system of linear equations in two variables**. A **solution** to the system is an ordered pair that satisfies both equations simultaneously. This activity explores solving linear systems using three methods (intersection, substitution, and elimination) and the possible numbers of solutions that can arise from linear systems in two variables.

Intersection Method. Both equations are graphed on the same coordinate plane and the solution is determined visually from the graph as the point of intersection of the two lines.

For example: *Solve* $\begin{array}{rcl} y & = & 3x + 5 \\ 2x + y & = & -5 \end{array}$ *by the intersection method.*

1.   Graph both linear equations on the same coordinate plane.

2.   Write the point of intersection of the two lines as an ordered pair. This is the solution to the system of equations.

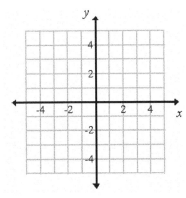

Substitution Method: One linear equation can be solved for a variable, and then replace that variable in the second equation with the resulting expression. The result will be one equation with one variable, which can easily be solved.

For example: *Solve* $\begin{array}{rcl} 2x + 3y & = & 4 \\ -6x + 2y & = & 10 \end{array}$ *by the substitution method.*

Solve the first equation for y to get $y = \frac{-2}{3}x + \frac{4}{3}$. Then replace y in the second equation with $\frac{-2}{3}x + \frac{4}{3}$ to obtain an equation in terms of x.

3.   Determine the equation when the expression $\frac{-2}{3}x + \frac{4}{3}$ replaces y in the equation $-6x + 2y = 10$.

4.   Solve this equation for x.

5.   Then substitute this value of x into either equation to solve for y. Report your solution as an ordered pair.

Elimination Method: This method requires algebra. The idea is to multiply one or both equations as needed to create coefficients of one variable that have opposite signs (such as $-3x$ and $3x$, or $5y$ and $-5y$). Then we add the equations to eliminate a variable to produce one equation with one unknown, which can be solved easily for the variable. Then we substitute the value of the variable into one of the original equations to solve for the other variable.

For example: *Solve* $\begin{matrix} 3x+5y & = & 9 \\ 4x+2y & = & 14 \end{matrix}$ *by the elimination method.*

Let's eliminate the variable x. We multiply each side of the first equation by –4 and multiply each side of the second

equation by 3 so that the coefficients of x will have opposite signs to obtain $\begin{matrix} -12x-20y & = & -36 \\ 12x+6y & = & 42 \end{matrix}$.

6.   Add the two left-hand sides of the equations and the two right-hand sides of the equations to obtain an equation in terms of y. You should get $-14y = 6$.

7.   Solve for y.

8.   Select one of the original equations, and replace y with the value from the previous question to obtain an equation in terms of one variable x. Solve for x.

9.   Write the solution to the system of linear equations as an ordered pair.

10. Apply the method of elimination to the same system of linear equations, but multiply each equation by appropriate numbers so that you can eliminate the variable y.

11. Solve $\begin{matrix} -2x+3y & = & 4 \\ 3x-4y & = & -5 \end{matrix}$ using all three methods.

12. Discuss advantages and disadvantages of the graphing method.

13. Discuss advantages and disadvantages of the substitution method.

14. Discuss advantages and disadvantages of the elimination method.

15. Why is it instructionally valuable to teach all three methods?

16. The perimeter of a rectangular swimming pool is 78 feet. The width is three more than twice the length. Find the dimensions of the pool.

ACTIVITY 7.2.2   Solving a Linear System and Interpreting the Results

Material: sharp pencil   Name _____

A system of two linear equations may have one solution (the two lines intersect at one point), no solution (the two distinct lines are parallel), or infinitely many solutions (the two lines are identical). In this activity, you will solve a system of two linear equations using the method of substitution. This will transform the two linear equations into one linear equation in one variable, which in turn can be transformed into an equation of the form $x = a$, $a = a$, or $a = b$, where x is a variable and a and b are different numbers. For example, we can solve the system of linear equations $2y = x + 10$ and $y = -3x - 2$ using the method of substitution by solving for x. We replace y with $-3x - 2$ in the first equation to obtain $2(-3x - 2) = x + 10$, which in turn is transformed into the equation $x = -2$. You will learn how to interpret the resulting equations $x = a$, $a = a$, or $a = b$.

1. There is one solution to each system (I, II, and III) of linear equations below. Solve each system using the method of substitution by solving for x. Then explain what each system had in common when you solved them.

    System I $y = 2x + 5$ and $y = -4x - 1$

    System II $2y = x + 8$ and $y = 3x + 4$

    System III $y = x + 1$ and $y = -3x + 5$

2. There are infinitely many solutions to each system (I, II, and III) of linear equations below. Solve each system using the method of substitution. Then explain what each system had in common when you solved them.

    System I $x - y = 10$ and $2x = 2y + 20$

    System II $4y = 3x + 20$ and $-15x + 20y = 100$

    System III $x - 3y = 6$ and $15y + 30 = 5x$

3. There are no solutions to each system (I, II, and III) of linear equations below. Solve each system using the method of substitution. Then explain what each system had in common when you solved them.

    System I $-3x + y = 7$ and $y = 3x - 1$

    System II $10y = 6x + 2$ and $5y - 3x = -8$

    System III $y = 12x$ and $2y = 24x + 10$

4. As you know from Questions 1-3, a system of two linear equations can be turned into an equivalent equation having one the following three forms: $x = a$, $a = a$, or $a = b$, where x is a variable and a and b are different numbers. Suppose two linear equations are turned into the given equivalent equation. Tell if the system of linear equations has one solution, infinitely many solutions, or no solution.

    a. $x = 10$ **b.** $-3 = 4$

    c. $3 = 5$ **d.** $-1 = -1$

    e. $7 = 7$ **f.** $x = 1$

ACTIVITY 7.3.1 Algebra Tiles

Material: sharp pencil and algebra tiles (p. A8)   Name _____

Algebra tiles give students a concrete model of variable and variable expressions. They provide a direct link to the use of algebra for solving one-step equations such as $x + 3 = 5$ and $3n = 6$, as well as two-step equations such as $2n - 1 = 5$. Elementary students typically learn how to solve one-step and two-step equations involving whole numbers. Later, these skills provide the foundation for working with equations involving integers, fractions, and decimals. Like the chip model for integers, we use two colors to distinguish positive and negative quantities. The following figure gives various examples for algebra tiles.

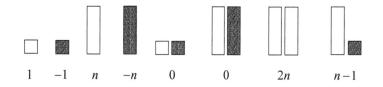

1   −1 n $-n$ 0   0 $2n$ $n-1$

The 1-tile (one tile), −1-tile (negative one tile), n-tile (n tile), and $-n$-tile (negative n tile) form the building blocks of expressions. Note that $n - 1$ is the same as $n + (-1)$ by adding the opposite, so we represent the expression $n - 1$ with one n-tile and one −1-tile. Similarly, $3 - 2n$ is the same as $3 + (-2n)$, so we represent $3 - 2n$ with three 1-tiles and two $-n$-tiles.

1.   Write the algebraic expression for the tiles. Use n for the variable.

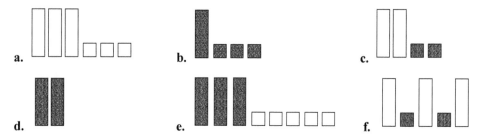

2.   Sketch the tile representation for each algebraic expression.
      a. $-n + 2$

      b. $-2n - 3$

      c. $2n + 1$

      d. $-2n + 1$

      e. $-3n - 7$

3.   Write the simplest algebraic expression for each collection of algebra tiles.

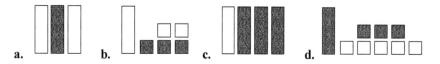

4. Discuss how you could use algebra tiles to model the expression $3(2n - 4)$.

5. Model each expression with algebra tiles. Then *add* the expressions by combining like tiles and removing any zero pairs. Then write the expression represented by the final collection of tiles.

   a. $n + 2$ and $2n - 4$

   b. $5 - 2n$ and $2n - 3$

   c. $3n - 1$ and $-2n + 4$

The general strategy in solving equations using algebra tiles is to make the same changes to both sides of the mat—much like the same changes to both sides of a balanced scale. The expression on each side of the mat changes, but the expressions represented on both sides of the mat are still equivalent. The goal is to transform the original equation into a form such as $n = 5$, because this type of equation reveals the solution.

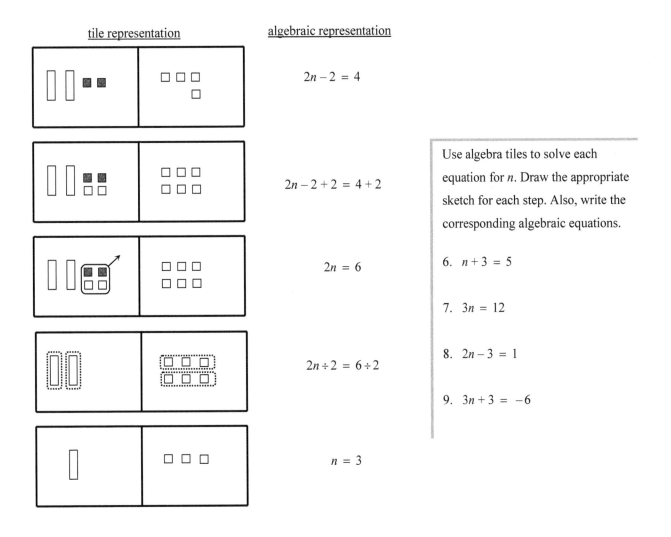

tile representation   algebraic representation

$2n - 2 = 4$

$2n - 2 + 2 = 4 + 2$

$2n = 6$

$2n \div 2 = 6 \div 2$

$n = 3$

Use algebra tiles to solve each equation for *n*. Draw the appropriate sketch for each step. Also, write the corresponding algebraic equations.

6. $n + 3 = 5$

7. $3n = 12$

8. $2n - 3 = 1$

9. $3n + 3 = -6$

ACTIVITY 8.1.1 Representing Categorical Variables

Material: sharp pencil, protractor, and graph paper (p. A9)   Name _____

Statistics is the science of collecting, organizing, visualizing, summarizing, and analyzing data. A **variable** is a characteristic that varies from one object to another. A **numerical variable** (also known as a *quantitative variable*) has values that represent counts or measurements. The values answer the question "How many?" or "How much?" Some examples of numerical variables are the number of pets in a home (because the value is determined by counting) and the weight of a person (because it must be measured with a device). A **categorical variable** (also known as a *qualitative variable*) has values that are labels or categories. The value of a categorical variable answers the question "Which category does the object belong to?" Some examples of categorical variables are hair color (with values such as black, brown, or red) or favorite type of movie (with values such as comedy, drama, horror, or adventure). In this activity, we will focus on ways to represent categorical data using a table (tally table, frequency table, and relative frequency table), pie chart, and bar graph.

1. Suppose the following table records the results of a survey in which students were asked their favorite primary color (red, blue, or green).

   a. What is the variable in this survey?

   b. Why is it a variable?

   c. Why is it a categorical variable?

   d. Check your answers with another student.

student	favorite color
Nikki	blue
Ziad	blue
Owen	red
Kevin	green
Matthew	blue
Olivia	green
Jack	blue
Cameron	blue

2. Complete the tables that summarize the data in the previous problem.

**Table 1.
tally table**

color	tally
blue	
red	
green	‖

**Table 2.
frequency table**

color	frequency
blue	
red	
green	2

**Table 3.
relative frequency table**

color	relative frequency
blue	
red	
green	25.00%

3. The frequency tables serve as an aid to display the data with a pie chart as shown. In a pie chart, each category is represented by a sector (slice) of the circle. The angle of a sector is called a *central angle*. Compute the measure of each central angle using the data in

   a. Table 2.

   b. Table 3.

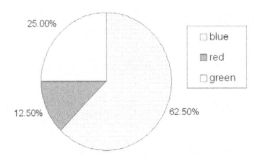

4. A student records the results of a brief survey of some classmates in the table shown.

    Calculate the central angle for each sector of the pie graph for the three categories.

    a. apples

    b. pears

    c. grapes

favorite fruit	frequency
apples	14
pears	10
grapes	6

5. Mrs. Smith asked her sixth-grade students which subject they liked. The actual results are shown in the table.

    a. Summarize the results in a tally table.

    b. Summarize the results in frequency table.

    c. Display the data using a pie chart. Be sure to include a title and legend.

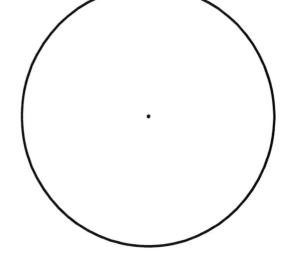

student	boy or girl	favorite subject
Nikki	g	art
Ziad	b	math
Owen	b	math
Kevin	b	P.E.
Matthew	b	P.E.
Olivia	g	P.E.
Jack	b	P.E.
Cameron	b	art
Turner	b	P.E.
Nicole	g	P.E.
Claire	g	P.E.
Madeline	g	P.E.
Devon	b	P.E.
Cale	g	math
Will	b	P.E.
Kai	g	science
Veronica	g	art
Trevor	b	P.E.
Ryan	b	P.E.
Annie	g	art
Rachael	g	English
Carolyn	g	English
Emily	g	art
Billy	b	P.E.

Source: Mrs. Smith, sixth grade teacher

6. Find an example of a pie chart from a newspaper, magazine, or non-academic Internet website.

    a. Is the title accurate and interpretative?

    b. Are the categories clearly labeled?

    c. Are the slices of the pie proportional to the percentages they represent?

7. A *two-way table* is a table that shows the association of two categorical variables.

    a. Complete the two-way table that summarizes the data in Problem 5.

    b. What are the two categorical variables?

	art	English	math	P.E.	science
boys			2		0
girls			1		1

8. The data in a two-way table can be displayed using a bar graph.

    Graphs A and B show different representations of the data in the two-way table in problem 7.

    a. Which bar graph highlights differences between the boys and girls?

    b. Which bar graph highlights differences among the subjects?

Graph A   **Graph B**

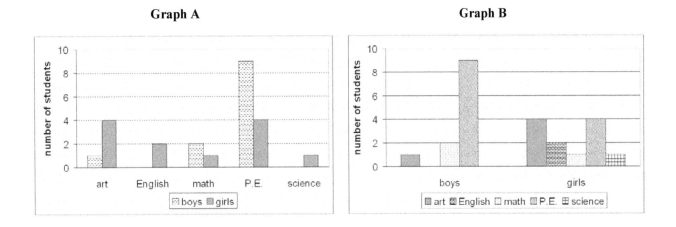

9. The following two-way table provides a glimpse of the latest data available on cybercrime. *Cyber attacks* are crimes against computer systems, such as computer virus attacks or denial of service attacks. *Cyber thefts* are crimes where a computer is used to steal money or items of value. *Other thefts* include unwelcome nuisances such as spyware, adware, hacking, or theft of information.

number of employees in the businesses	Cyber attack	Cyber theft	Other thefts
2-24	44%	8%	15%
25-99	51%	7%	17%
100-999	60%	9%	24%
1000 or more	72%	20%	36%

Source: U.S. Department of Justice, Bureau of Justice Statistics, 2008 NCJ 221943

Graphs A and B show two different ways to display the data using a bar chart.

    a. Which graph highlights differences among the number of employees in the businesses?

    b. Which graph highlights differences among the types of attacks?

    c. Write a descriptive title for graph A.

    d. Write a descriptive title for graph B.

    e. Share your answers with other students in the class.

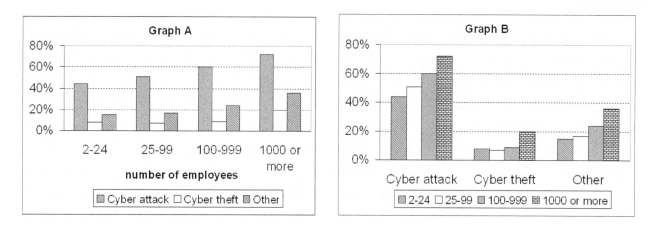

10. The following table shows the most recent Oscar winners for best actor and actress. The graph paper is on page A9.

   a. Summarize the ages of the actors with a tally chart using the groupings 20-29, 30-39, 40-49, 50-59, and 60-69.

   b. Display the data for the ages of the actors using a bar chart .Be sure to include a title and legend.

   c. Summarize the ages of the actresses with a tally chart using the groupings 20-29, 30-39, 40-49, 50-59, and 60-69.

   d. Display the data for the ages of the actresses using a bar chart. Include a title and legend.

   e. Create a two-way table for the data.

   f. Use the two-way table to create a bar graph that highlights differences between the age groups. Include a legend and descriptive title for the graph.

   g. Use the two-way table to create a bar graph that highlights gender differences. Include a legend and descriptive title for the graph.

Oscar Winners for Best Actor and Actress	
Actor, Movie, Age	**Actress, Movie, Age**
Colin Firth, The King's Speech, 50	Natalie Portman, Black Swan, 29
Jeff Bridges, Crazy Heart, 60	Sandra Bullock, The Blind Side, 45
Sean Penn, Milk, 48	Kate Winslet, The Reader, 33
Daniel Day-Lewis, There Will Be Blood, 50	Marion Cotillard, La Vie en rose, 32
Forest Whitaker, The Last King of Scotland, 45	Helen Mirren, The Queen, 61
Philip Seymour Hoffman, Capote, 38	Reese Witherspoon, Walk the Line, 29
Jamie Foxx, Ray, 37	Hilary Swank, Million Dollar Baby, 30
Sean Penn, Mystic River, 43	Charlize Theron, Monster, 28
Adrien Brody, The Pianist, 29	Nicole Kidman, The Hours, 35
Denzel Washington, Training Day, 47	Halle Berry, Monster's Ball, 35
Russell Crowe, Gladiator, 36	Julia Roberts, Erin Brockovich, 33
Kevin Spacey, American Beauty, 40	Hilary Swank, Boys Don't Cry, 25
Roberto Benigni, Life is Beautiful, 46	Gwyneth Paltrow, Shakespeare in Love, 26
Jack Nicholson, As Good as it Gets, 60	Helen Hunt, As Good as it Gets, 34
Geoffrey Rush, Shine, 45	Frances McDormand, Fargo, 39
Nicholas Cage, Leaving Las Vegas, 32	Susan Sarandon, Dead Man Walking, 49
Tom Hanks, Forrest Gump, 38	Jessica Lange, Blue Sky, 45
Tom Hanks, Philadelphia, 37	Holly Hunter, The Piano, 36
Al Pacino, Scent of a Woman, 52	Emma Thompson, Howards End, 33
Anthony Hopkins, The Silence of the Lambs, 54	Jodie Foster, The Silence of the Lambs, 29

ACTIVITY 8.2.1   Exploring Measures of Center

Material: sharp pencil and 15 pennies   Name _____

Maria is a middle school student elected to serve as the president of her class. Part of her job is to represent the views and interests of her classmates. In mathematics, a *measure of center* is a representative number of a collection of numbers. It is one of the most important ways to summarize numerical data, especially in the sciences. Three common choices are mean, median, and mode.

1. Do the following.

    a. Make three stacks of pennies. The first stack has 3 pennies, the second stack has 4 pennies, and the third stack has 8 pennies. Rearrange the pennies so that each of the three stacks have the same number of pennies.

    b. Calculate the mean of the numbers 3, 4, and 8.

2. The teacher asked 12 students in the class how many children were in their family. The mean was 2.25 children. How many children were there in the families combined?

3. The mean of five scores is 80. Four of the scores are 75, 88, 90, and 68. What is the other score?

4. The mean of five quiz scores for Emily was 82. One of her scores was changed from a 68 to an 86 due to an error in recording the grades.

    a. Will the new mean increase or decrease?

    b. Calculate the new mean of the quiz scores.

5. A shopper made 7 trips to the store. The mean number of items purchased was 5.

    Suppose we know six of the numbers as shown in the dot plot. What is the other number?

    a.   b.   c.

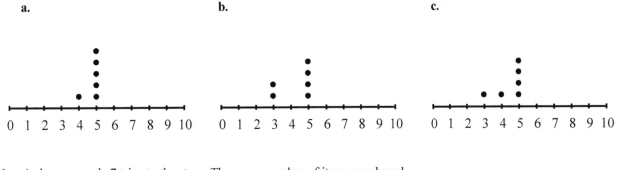

6. A shopper made 7 trips to the store. The mean number of items purchased was 5. Suppose we know five of the numbers as shown in the dot plot. What could the other two numbers be? Give three possible pairs. Check your answers with another student.

7. Suppose 6 gallons of blue paint costing a total of $15 are mixed with 4 gallons of yellow paint costing a total of $10. Find the mean cost for one gallon of the mixture.

8. A teacher asked her students "How many pets do you have?" After collecting the data, she determined that each student has an average of 1.4 pets. Some elementary students struggle with this answer because it has no physical counterpart in reality, since a student cannot have a fraction of a pet. How can you help your students make sense of this measure of center?

9. The median of the numbers 2, 4, 6, 8, 10 is 6. What is the mean of the numbers 4, 6, 2, 10, 8?

10. Compute the mean and median of the given numbers. Which measure of center is more sensitive to extreme values?
    a. 5, 7, 8, 12, 15
    b. 5, 7, 8, 12, 128

11. Salaries are often summarized with the median rather than the mean. Describe other situations in which the median would be more appropriate than the mean as a measure of center.

12. The stem-and-leaf plot displays quiz scores for a class.
    Some scores are 25, 25, 26, 30, and so on.
    a. How many quizzes were given?
    b. How many scores were in the 40s?
    c. Find the mean and median of the numbers.

stem	leaf 3|4 = 34
2	556
3	0136
4	122238
5	
6	02
7	

13. Pam purchased a bag of chips for $2.40, a soda for $0.80, and a sandwich for $3.70. If each item cost the same, how much would each item cost?

14. Give an example of a collection of 7 numbers such that the mean is 12 and the median is 10.

15. Two classes raised money for a fundraiser. Which class performed better?

	no. of students	amount raised
2nd graders	15	$219
3rd graders	23	$294.40

16. Your teacher gives you a collection of 5 numbers such that the mean is 15. How could you find a collection of numbers such that the mean is 45? Apply and check your strategy with the collection of numbers 4, 6, 17, 22, 26.

17. Your teacher gives you a collection of 5 numbers such that the mean is 15. How could you find a collection of numbers such that the mean is 18? Apply and check your strategy with the collection of numbers 4, 6, 17, 22, 26.

ACTIVITY 8.2.2   Grasping Variation

Material: sharp pencil   Name _____

When we collect numerical data $x_1, x_2, \ldots, x_n$, there will be variation in the values. In this activity, we will use the dot plot to visualize variation and explore the mean absolute deviation (MAD) as a measure of variation.

1. The dot plots display the waiting times (in minutes) for five customers at two banks.
    a. How are the two data sets alike?
    b. How are the data sets different?
    c. Based on the dot plots, which bank would you prefer? Justify your answer.

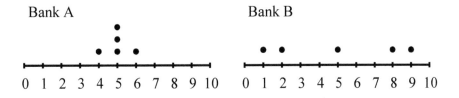

2. The data sets show the lifespan, in hours, of two brands of batteries. Calculate the mean, median, and mode for each data set. Then draw a line plot (dot plot) for each data set. Which battery do you prefer?

    Brand A: 5.3, 5.5, 6, 6.5, 6.5, 6.8, 7.4, 8   Brand B: 4, 4.7, 5, 6.5, 6.5, 7.7, 7.9, 9.7

3. One way to measure the spread of the data x_1, x_2, and x_3 is to sum their *absolute deviations* from $\bar{x}$ and then divide by 3 to obtain the mean of the absolute deviations. This number is called the **mean absolute deviation (MAD)**:

    $$MAD = \frac{|x_1 - \bar{x}| + |x_2 - \bar{x}| + |x_3 - \bar{x}|}{3}.$$

    Note: In general, $MAD = \dfrac{|x_1 - \bar{x}| + |x_2 - \bar{x}| + \cdots + |x_n - \bar{x}|}{n}.$

    The MAD tells us how much each score, on the average, deviates from the mean. Find the MAD for each data set.
    a. 5, 6, 10
    b. 10, 16, 22, 8

As discussed in Section 8.2 of the textbook, "usual values" are values that you would expect to occur. Knowing the *minimum usual value* and the *maximum usual value* would be helpful in understanding variation. For example, suppose you know the minimum usual value is 150 and the maximum usual value is 180.

*   These values give you a sense of the variation of the data because you would know "typical values" are from 150 to 180.

*   These values help you identify whether a score is unusual or usual. For example, 110 would be considered an unusually low value, 160 would be considered a usual value, and 195 would be considered an unusually high value.

4. Let $\bar{x}$ and MAD represent the mean and mean absolute deviation of a collection of numbers.

   Refer to Section 8.2 in your textbook to complete the following rules of thumb for using the MAD to grasp variation:

   a. **Usual values** of the data belong to the interval [_____ , _____]

   b. The **minimum usual value** is _____ .

   c. The **maximum usual value** is _____ .

   d. If $x < \bar{x} - 2.5 \cdot \text{MAD}$, then x is considered an _____ value.

   e. If $\bar{x} + 2.5 \cdot \text{MAD} < x$, then x is a considered an _____ value.

5. A teacher gives a test. The mean is 75 and the mean absolute deviation is 6.

   a. Find the interval of usual values.

   b. Would a score of 55 be considered usual or unusual?

   c. Would a score of 64 be considered usual or unusual?

   d. Would a score of 87 be considered usual or unusual?

   e. Would a score of 95 be considered usual or unusual?

6. Find the interval of usual values for the data: 60, 66, 71, 83.

7. The organizer of the annual county fair said they need 1 employee for every 350 people at the fair to give directions, provide rides, and ensure the safety of the attendees. Past data shows that 6,000 people attend the fair daily on the average, with a mean absolute deviation of 400 people per day. How many employees should the manager schedule to work at the fair each day to give directions, provide rides, and ensure the safety of the attendees?

8. The teacher announces the average on the exam was 78.5 and the mean absolute deviation was 6.2. Jerry cannot remember what his friend scored, but he knows it was either a 58 or 85. What score do you think his friend received? Justify your answer.

ACTIVITY 8.3.1 Five-Number Summary and Box Plots

Material: sharp pencil and graph paper (p. A9)   Name _____

The **five-number summary** of a collection of numbers consists of five numbers:

- the lowest number (Min)
- the 25th percentile (Q_1), which is the first quartile (median of the lower half of the numbers)
- the 50th percentile (Q_2), which is the second quartile (median)
- the 75th percentile (Q_3), which is the third quartile (median of the upper half of the numbers)
- the highest number (Max)

Now we demonstrate how to find to find a five-number summary. First, find the median Q_2. Second, find the first quartile Q_1 (median of the lower half). Third, find the third quartile Q_3 (median of the upper half). Then find the highest and lowest values.

Data set 1: 56, 74, 76, 80, 86, 86, 87, 90, 95, 96, 99   Data set 2: 56, 74, 76, 80, 84, 86, 86, 90, 95, 96

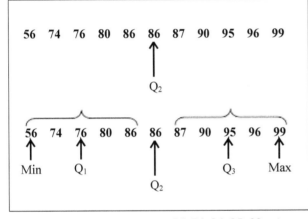

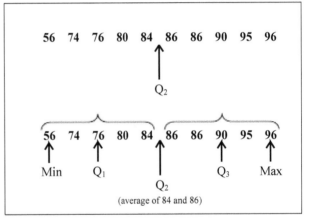

five number summary: 56, 76, 86, 95, 99   **five number summary: 56, 76, 85, 90, 96**

A **box plot** is a visual representation of a five-number summary. For example, a collection of numbers with the five-number summary 80, 88, 93, 99, 106 can be represented with a box plot as shown:

Now we illustrate how to compare two fictitious sets A and B using box plots.

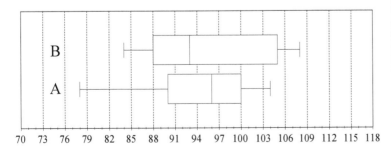

The interquartile range (IQR) is defined as the difference $Q_3 - Q_1$.
Set A: IQR $= Q_3 - Q_1 = 100 - 90 = 10$
Set B: IQR $= Q_3 - Q_1 = 105 - 88 = 17$

Rule of Thumb: The data with the larger IQR has more variation. This suggests that set B has more variation than set A due to the larger IQR.

You can plot the box plots on the graph paper on page A9.

1. The data shows the average NAEP mathematics scores for 4th grade students, based on the latest data from Digest of Education Statistics, 2010.

year	1990	1992	1996	2000	2003	2005	2007	2009
average	213	220	224	226	235	238	240	240

   a. Find the five-number summary for the data.

   b. Make a box plot for the data.

2. Find the five-number summary for the given box plot.

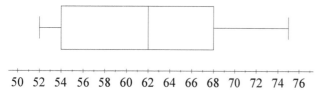

   50 52 54 56 58 60 62 64 66 68 70 72 74 76

3. The following box plot summarizes the public and private pupil/teacher ratios for grades K-12, for fall 1996 through fall 2008, based on the latest data from Digest of Education Statistics, 2010.

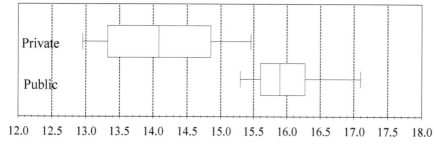

   12.0   12.5   13.0   13.5   14.0   14.5   15.0   15.5   16.0   16.5   17.0   17.5   18.0

   a. Which data set shows more variation in the pupil/teacher ratios? Explain your answer.

   b. Which educational group has a lower average of pupil/teacher ratio?

4. The data shows the average NAEP reading scores for 4th grade students, based on the latest data from Digest of Education Statistics, 2010.

year	1992	1994	1998	2000	2002	2003	2005	2007	2009
male	213	209	212	208	215	215	216	218	218
female	221	220	217	219	222	222	222	224	224

   a. Find the five number-summary for the males.

   b. Find the five-number summary for the females.

     (*Hint*: sort the data in nondecreasing order before finding the five-number summaries.)

   c. Make a box plot that compares both groups of students.

   d. Which group shows more variation in the average reading scores? Explain your answer.

   e. Which group has a higher average reading score?

5. The following table shows the most recent Oscar winners for best actor and actress.

   a. Find the five-number summary for the ages of the actors.

   b. Share your results with another student, make sure you have the same results.

   c. Find the five-number summary for the ages of the actresses.

   d. Share your results with another student, make sure you have the same results.

   e. Draw a box plot that compares both groups of winners.

   f. Write a brief paragraph about what the box plot reveals.

   g. Give your paragraph to another student to receive feedback on your writing.

Oscar Winners for Best Actor and Actress	
Actor, Movie, Age	**Actress, Movie, Age**
Colin Firth, The King's Speech, 50	Natalie Portman, Black Swan, 29
Jeff Bridges, Crazy Heart, 60	Sandra Bullock, The Blind Side, 45
Sean Penn, Milk, 48	Kate Winslet, The Reader, 33
Daniel Day-Lewis, There Will Be Blood, 50	Marion Cotillard, La Vie en rose, 32
Forest Whitaker, The Last King of Scotland, 45	Helen Mirren, The Queen, 61
Philip Seymour Hoffman, Capote, 38	Reese Witherspoon, Walk the Line, 29
Jamie Foxx, Ray, 37	Hilary Swank, Million Dollar Baby, 30
Sean Penn, Mystic River, 43	Charlize Theron, Monster, 28
Adrien Brody, The Pianist, 29	Nicole Kidman, The Hours, 35
Denzel Washington, Training Day, 47	Halle Berry, Monster's Ball, 35
Russell Crowe, Gladiator, 36	Julia Roberts, Erin Brockovich, 33
Kevin Spacey, American Beauty, 40	Hilary Swank, Boys Don't Cry, 25
Roberto Benigni, Life is Beautiful, 46	Gwyneth Paltrow, Shakespeare in Love, 26
Jack Nicholson, As Good as it Gets, 60	Helen Hunt, As Good as it Gets, 34
Geoffrey Rush, Shine, 45	Frances McDormand, Fargo, 39
Nicholas Cage, Leaving Las Vegas, 32	Susan Sarandon, Dead Man Walking, 49
Tom Hanks, Forrest Gump, 38	Jessica Lange, Blue Sky, 45
Tom Hanks, Philadelphia, 37	Holly Hunter, The Piano, 36
Al Pacino, Scent of a Woman, 52	Emma Thompson, Howards End, 33
Anthony Hopkins, The Silence of the Lambs, 54	Jodie Foster, The Silence of the Lambs, 29

6. The following table summarizes the average test scores of 4th graders from 36 countries in the areas of *geometric shapes and measures* (which includes "lines and angles, two- and three-dimensional shapes, and location and movement") and *data display* (which includes "reading and interpreting, and organizing and representing"), based on the latest data from Digest of Education Statistics, 2010. The tests were part of the 2007 Trends in International Mathematics and Sciences Study (TIMSS) which allows performance comparisons across participating countries.

   a. Which subject shows more variations in scores? Justify your answer.

   b. On the average, which of the two subjects resulted in higher scores?

subject	Minimum score	25th percentile	median	75th percentile	Maximum score
geometry and measurement	296	429	509	536	599
data display	307	414	512	533	585

ACTIVITY 9.1.1   Language of Probability and Equally Likely Outcomes

Material: sharp pencil   Name _____

Probability is a measure of the chance that a particular event will occur. We use probability to make informed decisions despite uncertainty. This activity explores vocabulary words used in probability and the probability rule for experiments with equally likely outcomes.

We use numbers from 0 to 1 to represent the likelihood that an event will occur.

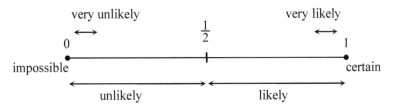

- An **experiment** is a procedure in which something happens, but the result cannot be predicted in advance. *For example, an experiment consists of randomly selecting a number from 1 to 10.*
- An **outcome** is the result of an experiment. *The outcome is a 7.*
- The **sample space** S is the set of all possible outcomes of an experiment. *$S = \{1, 2, 3, 4, 5, 6, 7, 8, 9, 10\}$*
- An **event** E is a collection of outcomes. *$E = \{5, 8\}$ is the event of selecting a 5 or 8.*

1. A spinner has outcomes red (R), blue (B), and green (G). Sketch a spinner with all sectors for R, B, and G that meets the qualitative description. Then show your sketches to another student for feedback.

    a. Green is very likely to occur.   **b.** Blue is likely to occur.   **c.** Red is unlikely to occur.

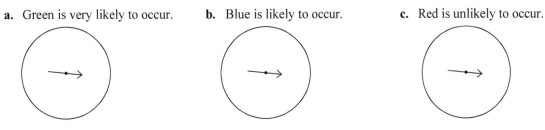

2. A bag has 20 marbles with only red, blue, and green marbles. Tell how many marbles of each color there could be in the bag that meets the qualitative description. Then show your answers to another student for feedback.

    a. Green is very likely to occur.   **b.** Blue is likely to occur.   **c.** Red is unlikely to occur.

3. An experiment consists of rolling a number cube and recording the number that shows on the top face. The net for the number cube is shown. Answer the following questions related to the experiment.

    a. List one possible outcome from this experiment.

    b. List the sample space S for this experiment.

    c. List the outcomes in the event E of obtaining an even number.

    d. Describe the event $E = \{3, 6\}$ in words.

    e. Describe the event $E = \{1, 2, 3\}$ in words.

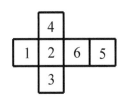

4. Suppose an event has probability 65% (or 0.65 or 65/100). This means that, ideally, if you repeat an experiment 100 times, then the event will occur 65 times. The teacher has a bag of marbles. An experiment consists of randomly picking one marble and then recording the color. Suppose the likelihood of drawing a green marble is 32%. If you repeat this experiment 140 times, how many green marbles would you expect to pick?

5. What does it mean when we say the chance it will rain is 73%?

Many experiments consist of *equally likely outcomes*. This means each outcome in the sample space has the same chance of occurring. For example,

- $S = \{T, H\}$ is the sample space for tossing a fair coin, where T = tails and H = heads. Then $P(T) = P(H) = 1/2$.
- A bag contains 8 green marbles, 3 blue marbles, and 1 yellow marble. An experiment consists of randomly picking one marble from the bag without peeking. Each marble has the same chance 1/12 of being picked because they have the same size and feel the same (but the color green is more likely to be picked).

If the sample space consists of equally likely outcomes, then the probability of event E is

$$P(E) = \frac{\text{the number of ways } E \text{ can occur}}{\text{the number of possible outcomes in the sample space}}.$$

6. A raffle has tickets consisting of two digits, from 00 to 99. One ticket will be picked and the winning number announced.
    a. What is the probability that 72 is picked?
    b. What is the probability 72, 73, 74, or 75 is picked?
    c. What are the chances that a ticket with digits from 12 to 30 is picked?
    d. What is the likelihood that both digits are identical?

7. A bag contains 8 green marbles, 3 blue marbles, and 1 yellow marble. An experiment consists of randomly picking one marble from the bag without peeking and noting the color.
    a. What is the probability of picking a blue marble?
    b. What is the probability of picking a green marble?

8. Bag A contains 18 blue marbles and 12 yellow marbles. Bag B contains 25 blue marbles and 20 yellow marbles. If you hoped to draw a blue marble, which bag would you choose to draw from? Explain your answer to another student.

9. An experiment consists of spinning the spinner shown and recording the number that the arrow points to.
    a. What is the probability that the arrow points to a 2?
    b. What are the chances that the arrow points to an even number?
    c. What is the likelihood that the arrow points to a 3 or a 4?
    d. What is the probability the arrow points to a 6?
    e. What is the probability the arrow points to a number less than 6?

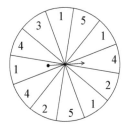

ACTIVITY 9.1.2 Equally Likely Outcomes, Again

Material: sharp pencil, straightedge, and protractor   Name _____

1. A number cube has numbers on the faces as shown in the net for the number cube. An experiment consists of rolling the number cube and noting the number on the top face of the cube. Each face has the same chance of showing.

    a. Explain why the probability of a 4 is 2/6.

    b. Find the probability that the outcome is a 3.

    c. Find the probability that the outcome is a 6.

    d. Find the probability that the outcome is a 1 or a 4.

    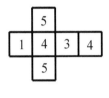

2. Cheryll and Marc buy raffle tickets consisting of 4 digits from 0000 to 9999. Cheryll has the ticket 7777 and Marc has the ticket 5721. Does Cheryll have a higher chance, lower chance, or the same chance as winning as Marc? Explain your answer to another student.

3. Two players play a game rolling two standard number cubes. The outcome is determined by the difference of the larger number and the smaller number. For example, if the two number cubes show a 2 and 6, then the outcome is 4 (6 − 2 = 4). Player A wins if the difference is 1, 3, or 4. Player B wins if the difference is 0, 2, or 5.

    a. Which player do you think has a higher chance of winning?

    b. Complete the table to show the possible outcomes of rolling the two number cubes.

    c. Find the probability that Player A wins.

    d. Find the probability that Player B wins.

    e. If they play 160 games, then how many games would you expect Player A to win?

    f. Which collection of outcomes would give the players the same chance of winning?

    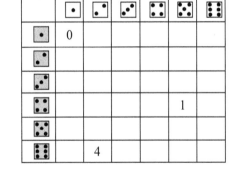

4. Any spinner can be split into 360 equal sized sectors, where each sector represents 1°. The spinner shown has outcomes A, B, and C. The experiment consists of spinning the spinner and recording the letter that the arrow points to.

    a. Explain how the formula for equally likely outcomes can be used to find the probability of each outcome.

    b. Find each probability:
    $P(A) =$
    $P(B) =$
    $P(C) =$

5. Draw a spinner with outcomes A, B, and C such that $P(A) = 45\%$, $P(B) = 35\%$, and $P(C) = 20\%$.

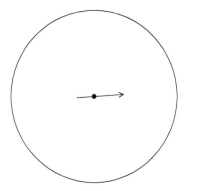

6. A standard deck of cards has 52 cards. There is an A (Ace), 2, 3, 4, 5, 6, 7, 8, 9, 10, J (Jack), Q (Queen), and K (King) in each of four different suits. The suits are hearts (♥), spades (♠), clubs (♣), and diamonds (♦). The diamonds and hearts are red while the clubs and spades are black. An experiment consists of randomly drawing one card from the deck and recording the card.

   a. Are the outcomes equally likely?

   b. What is the probability of drawing an Ace of spades?

   c. What is the probability of drawing an Ace?

   d. What is the probability of drawing a red card?

   e. What is the probability of drawing a card valued from 5 to 9?

7. An experiment consists of rolling two number cubes with 1, 2, 3, 4, 5, and 6 and adding the two numbers that appear on the top face.

   a. Complete the table of sums.

sum	⚀	⚁	⚂	⚃	⚄	⚅
⚀	2					
⚁						
⚂						
⚃				9		
⚄						
⚅	8					

   b. How many ways can you roll a 7?

   c. What is the probability of rolling a 7?

   d. Complete the table to determine the probability of each outcome.

outcome	2	3	4	5	6	7	8	9	10	11	12
probability											

   e. What is the probability a 5 is rolled?

   f. What is the probability a 5 is not rolled?

ACTIVITY 9.1.3 Basics of Experimental Probability

Material: sharp pencil, coin, and graph paper (pp. A10 and A11)   Name _____

In this activity, we will learn about the short-term and long-term behavior of experimental results. Let's consider an experiment that consists of tossing a fair coin and recording the outcome of heads (H) or tails (T). Suppose you toss the coin 10 times, with the following results: H H T H T H T T T T. Based on these results, the experimental probability of tails is 6/10, or 0.60. Another student may toss the coin 10 times and obtain the following results: H T H T H H H T T H. Based on these results, the experimental probability of tails is 4/10, or 0.40. **Experimental probability** is a quotient f/n, where f is the number of times you observed that the event occurred and n is the number of times the experiment was repeated.

1. Work in a group of three students. Flip a coin 100 times. One student should flip the coin, another student should count the number of flips, and another student should record the data.

    a. Complete the table for your group. These results will vary among the groups of students.

n, number of flips	f, number of tails	f/n, decimal (nearest 0.01)
5		
10		
15		
20		
25		
30		
35		
40		
45		
50		
60		
70		
80		
90		
100		

    b. Graph your results from part (a) using the graph paper in A10, connecting the points with line segments.

    c. Each group should write their results of f for $n = 25$ on the chalk board or white board. Label the results Group 1, Group 2, and so on. Record the combined results in the table. Enter the entry for Group 1 in the first row. Enter the combined results for Groups 1 and 2 in the second row, and so on. Each group should obtain the same results in this table.

n, number of flips	f, number of tails	f/n, decimal (nearest 0.01)
25		
50		
75		
100		
125		
150		
175		
200		
225		
250		

    d. Graph your results from part (c) using the graph paper in A11, connecting the points with line segments.

e. Do the results seem to match the theoretical probability of tails (0.5) for small values of *n*, say *n* = 5, 10, 15, or 20? Small values of *n* represent "short-term behavior."

f. Do the results seem to match the theoretical probability of tails (0.5) for higher values of *n*, say *n* = 350, 400, 450, or 500? Large values of *n* represent "long-term behavior."

g. Which part of the graph tends to depart from the expected probability of tails (0.5), the graph for smaller values of *n* or the graph for larger values of *n*?

The **Law of Large Numbers** (**LLN**), proven in the early 1700s, states that the fraction *f*/*n* of times an event *E* occurs when an experiment is repeated *n* times approaches a fixed number as *n* increases. That fixed number is the theoretical probability $P(E)$ that *E* occurs.

2. An experiment consists of drawing two marbles from a bag. *E* is the event both marbles have the same color. The graph shows experimental results of the probability of *E* for various values of *n*. Examine the graph to estimate $P(E)$.

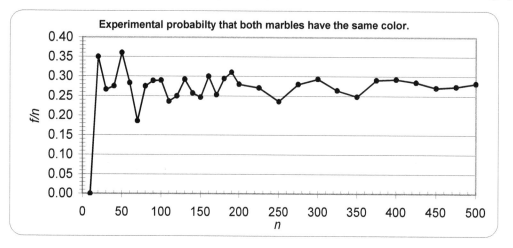

3. The table shows data for Taylor, Richard, and Sher.

name	number of times the event *E* occurred	number of times *n* the experiment was repeated
Taylor	70	135
Richard	145	300
Sher	280	500

a. For each student, determine the experimental probability that *E* will occur.

b. Which experimental probability seems like the best estimate of $P(E)$?

c. How could these results be combined in some way to obtain a possibly better estimate $P(E)$?

4. A bag contains 50 blue (B), 20 green (G), and 10 yellow (Y) marbles. An experiment consists of randomly picking one marble from the bag, recording the color, and returning the marble to the bag. When we randomly pick a marble from the bag, we do not know the color of the marble in advance. For that reason, we cannot predict the next term in the sequence BGYBGYBGY... where B represents a blue marble was picked, G represents a green marble was picked, and Y represents a yellow marble was picked.

a. What aspect of the sequence is predictable if repeat this experiment a large number of times, say 3000 times?

b. What aspect of the sequence is unpredictable, besides the next outcome, if we repeat the experiment a small number of times, say 20 times?

ACTIVITY 9.1.4 Experimental Probability, Again

Material: sharp pencil, tack, and paper clip   Name _____

As a group you will be performing an experiment. The experiment consists of spinning a spinner and gathering data. Then you will compare experimental and theoretical results.

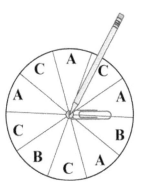

1.   The spinner shown has outcomes A, B, and C.

     Determine the theoretical probability that the spinner points to A.

2.   Use a paper clip and a pencil as a spinner as shown. Flick the paper clip with your fingers to spin the paper clip. Record your results in the table for various values of n. Once you enter the data for $n = 10$, then you should acquire new data for $n = 20$ rather than re-use the data for $n = 10$, and so on. Students can pool fresh results for higher values of n. Discuss how the class could work together to complete the table. Then complete the table.

n, number of spins	f, number of times the spinner points to A	f/n, experimental probability (nearest 0.001)
10		
15		
20		
25		
30		
35		
40		
45		
50		
100		
125		
150		
175		
200		

3.   Do the results seem to match the theoretical probability that the spinner points to A (0.4) for small values of n, say $n = 10, 15, 20,$ or 25? Small values of n represent "short-term behavior."

4.   Do the results seem to match the theoretical probability that the spinner points to A (0.4) for higher values of n, say $n = 125, 150, 175,$ or 200? Large values of n represent "long-term behavior."

5.   Compare the experimental probabilities in the table to the theoretical probability in Problem 1.

6.   Write a brief paragraph of what you learned in this activity.

As a group you will be performing another experiment. The experiment involves tossing a tack and gathering data about how the tack lands.

7.   What are the possible outcomes for this experiment of tossing a tack?
8.   Estimate the probability of each outcome. Explain your reasoning.

Logically there are only two possible outcomes of this experiment. One outcome occurs when the tack lands point down (D). The other outcome occurs when the tacks lands point up (U). A statement of D's and U's such as DDUD represents 4 experiments where the tack landed point down on the first, second, and fourth experiment, and landed point up on the third experiment.

        point down (D)   point up (U)

9.   Toss the tack *n* times for various values of *n* as indicated in the table. Once you enter the data for *n* = 10 tosses, then you should acquire new data for *n* = 20 rather than re-use the data for *n* = 10, and so on. Students can pool fresh results for higher values of *n* to spread the workload. Discuss how the class could work together to complete the table. Then complete the table.

n, number of tosses	number times the tack lands point		*experimental probability* (nearest 0.01)	
	down	**up**	*D*	*U*
10				
20				
30				
40				
50				
60				
70				
80				
90				
100				
120				
140				
160				
180				
200				

10. Do the experimental probabilities for D and U seem to be approaching specific numbers?
11. Based on the data available, estimate the theoretical probability of each outcome D and U.

ACTIVITY 9.2.1 Probability Involving ORs

Material: sharp pencil   Name _____

This activity focuses on applying formulas for calculating probabilities for events "*A* or *B*" and "*A* and *B*."

Mutually exclusive events are ones that cannot occur at the same time. For example, consider an experiment that consists of rolling a 1-6 number cube and recording the number that appears on the top face of the number cube. The events $A = \{1, 2\}$ and $B = \{4, 5, 6\}$ cannot occur at the same time because they have no common outcomes. In addition,

- A or $B = \{1, 2, 4, 5, 6\}$ and $P(A$ or $B) = \frac{5}{6}$

- $P(A) = \frac{2}{6}$, $P(B) = \frac{3}{6}$, and $P(A$ or $B) = P(A) + P(B)$.

1. An experiment consists of randomly picking one card from a standard deck of cards.
    a. What is the probability that the card is a 7?
    b. What is the probability that the card is a King?
    c. A is the event the card is a 7 and B is the event the card is a King. Are these events mutually exclusive?
    d. What is the probability that the card is a 7 or King?
    e. What is the probability that the card is a 5?
    f. What is the probability that the card is a 9?
    g. A is the event the card is 5 and B is the event the card is a 9. Are these events mutually exclusive?
    h. What is the probability that the card is a 5 or 9?

2. A bag contains 5 red marbles, 8 green marbles and 2 blue marbles. An experiment consists of randomly picking a marble and noting the color.
    a. What is the probability of picking a red marble?
    b. What is the probability of picking a blue marble?
    c. A is the event the marble is red and B is the event the marble is blue. Are these events mutually exclusive?
    d. What is the probability of picking a red or blue marble?

Suppose an experiment consists of spinning the spinner shown. Let A be the event the arrow points to a prime number and let B be the event the arrow points to an even number. Then events $A = \{2, 3, 5\}$ and $B = \{2, 6\}$ can occur at the same time because of the common outcome of 2. It turns out that if A and B have a common outcome, then $P(A$ or $B) \neq P(A) + P(B)$. Here's why for this example:

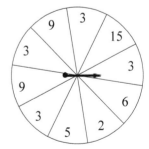

- A or $B = \{2, 3, 5, 6\}$ and $P(A$ or $B) = \frac{7}{10}$;

- $P(A) = \frac{6}{10}$, $P(B) = \frac{2}{10}$, so $P(A) + P(B) = \frac{8}{10}$;

- So $P(A$ or $B) \neq P(A) + P(B)$ (because we counted the common outcomes twice).

Event A and $B = \{2\}$ and $P(A$ and $B) = \frac{1}{10}$. Then $P(A$ or $B) = P(A) + P(B) - P(A$ and $B)$, because $\frac{7}{10} = \frac{6}{10} + \frac{2}{10} - \frac{1}{10}$. The formula $P(A$ or $B) = P(A) + P(B) - P(A$ and $B)$ holds for any two events A and B.

3. An experiment consists of randomly picking one card from a standard deck of cards.

   a. What is the probability the card is a 7?

   b. What is the probability the card is a heart?

   c. What is the probability the card is a 7 or a heart?

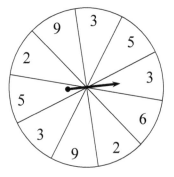

4. Suppose an experiment consists of spinning the spinner shown.

   a. What is the probability the arrow points to a prime number?

   b. What is the probability the arrow points to an even number?

   c. What is the probability the arrow points to a prime number or even number?

5. Students in a class were asked to name their favorite subject among math, art, and spelling. The following table summarizes the results of the survey.

	math	art	spelling
Male	10	6	4
Female	12	5	6

   a. What is the probability that a randomly selected student is a male?

   b. What is the probability that a randomly selected student prefers art?

   c. What is the probability that a randomly selected student is a female or prefers spelling?

6. A bag contains 8 red marbles, 12 green marbles and 5 blue marbles. An experiment consists of randomly picking a marble and noting the color.

   a. What is the probability of picking a red marble?

   b. What is the probability of picking a blue marble?

   c. What is the probability of picking a red or blue marble?

7. A game involves rolling two 1-6 number cubes and adding the results of the numbers appearing on the top face of the cubes. What is the probability the player rolls an 8 or doubles?

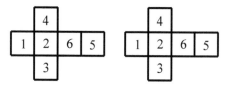

ACTIVITY 9.2.2 Probability Involving ANDs for Independent Events.

Material: sharp pencil   Name _____

This activity focuses on applying formulas for calculating probabilities for events "*A* and *B*" when *A* and *B* are independent events. When we say event *A* and event *B* are **independent events**, we mean that the occurrence of one of the events does not affect the probability of the second event. For example, suppose a bag contains 5 green marbles and 7 yellow marbles. As you know, the probability of randomly picking a green marble is 5/12 while the probability of randomly picking a yellow marble is 7/12. Benny randomly picks a marble, and it is green. He returns it to the bag. Then Holly randomly picks a marble, and it is yellow. The fact that a green marble was picked first does not affect the probability of choosing a yellow marble because the probability is still 7/12. When we say the two events *A* and *B* are **dependent events**, we mean the fact that *A* occurs does affect the probability the event *B* will occur. For example, suppose Benny picks a green marble but does not return it to the bag. Now Holly randomly picks a marble, and it is yellow. The probability Holly randomly picks a yellow marble is 7/11, because there were 7 yellow marbles and 11 marbles in the bag. So the fact that a green marble was picked first affects the probability the second marble is yellow since the probability changed from 7/12 to 7/11.

Drawing several cards with replacement, tossing a coin repeatedly, or spinning several spinners involve independent events and therefore multiplying probabilities (the details are in the textbook). A bag contains 5 green marbles and 7 yellow marbles. Two marbles are randomly picked, one at a time, with replacement. What is the probability the first marble is green and the second marble is yellow? The answer is 24%, because $P(GY) = P(G) \cdot P(Y) = \frac{5}{12} \cdot \frac{7}{12} \approx 24\%$.

1. A bag contains 3 green marbles, 5 yellow marbles, and 7 red marbles. Two marbles are randomly picked, one at a time, with replacement. What is the probability the first marble is yellow and the second marble is green?

2. Maria selects two cards from a standard deck of cards, one at a time, with replacement. What is the probability the first card is a 7 and the second card is a 5?

3. A jar contains 6 pennies, 8 nickels, and 12 dimes. Gary randomly picks a coin, and then returns it to the jar. Then Christie randomly picks a coin. What is the probability the first coin is a penny and the second coin is a dime?

4. Nancy rolls a 1-6 number cube three times. What is the probability that she rolls three sixes?

5. Nancy rolls three 1-6 number cubes. What is the probability that each number cube is a six?

6. The Math Club decides to raise money using lotteries in September and October. The probability of winning the first lottery is $\frac{65}{250}$ and the probability of winning the second lottery is $\frac{80}{365}$. What is the probability that a person wins both lotteries?

7. A box contains 10 blue marbles, 5 red marbles, and 3 green marbles. A marble is drawn from the box, with replacement, until a green marble is picked. Determine the probability of each event.

    a. One draw is needed.

    b. Two draws are needed.

    c. Three draws are needed.

ACTIVITY 9.2.3 Tree Diagrams

Material: sharp pencil   Name _____

It is easier to analyze complex probability situations using a tree diagram when they involve several stages, such as randomly drawing several cards, randomly picking several marbles, spinning two spinners, or rolling several die. These types of experiments are **multistage** experiments because they involve more than one stage. The main idea of a tree diagram is to represent the outcomes using branches (line segments) and represent probabilities using numerical labels on the branches. In this activity, you will learn how to use tree diagrams to analyze and solve probability problems.

1. A jar contains 5 green (G) marbles, 2 yellow (Y) marbles, and 3 red (R) marbles. An experiment consists of randomly drawing two marbles, one at a time, with replacement. This means that the first marble is replaced before the second marble is drawn. The line segments in the tree represent possible outcomes, and the weights along the line segments represent probabilities.

    - One possible outcome of the first draw is a G marble. The probability of drawing a G marble is 5/10.
    - Given the first marble is G, one possible outcome of the second draw is a Y marble. Given the first marble is G, the probability of drawing a Y marble is 2/10.
    - One outcome of this experiment is drawing G and Y marbles, in that order. The outcome GY represents the first marble is green *and* the second marble is yellow, given the first marble is green. The probability of an outcome is obtained by multiplying the probabilities along the branches. For example, $P(GY) = \frac{5}{10} \cdot \frac{2}{10} = \frac{10}{100}$.

    a. Complete the tree diagram for this experiment by listing the outcomes and probabilities. Check your answers with another student.

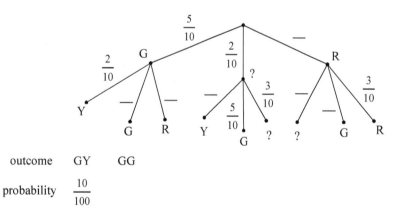

    b. What is the probability of drawing two green marbles?
    c. What is the probability the first marble is yellow and the second marble is red?
    d. What is the probability the both marbles have the same color?
    e. What is the probability the second marble is yellow, given the first marble is red?
    f. What is the probability the second marble is green, given the first marble is green?
    g. What is the probability the first marble is yellow and the second marble is green?

2. A jar contains 5 green (G) marbles, 2 yellow (Y) marbles, and 3 red (R) marbles. An experiment consists of randomly drawing two marbles, one at a time, without replacement. This means that the first marble is not replaced before the second marble is drawn. The line segments in the tree represent possible outcomes, and the weights along the line segments represent probabilities.

- One possible outcome of the first draw is a G marble. The probability of drawing a G marble is 5/10.
- Given the first marble is G, one possible outcome of the second draw is a Y marble. Given the first marble is G, the probability of drawing a Y marble is 2/9.
- One outcome of this experiment is drawing G and Y marbles, in that order. The outcome GY represents the first marble is green *and* the second marble is yellow, given the first marble is green. The probability of an outcome is obtained by multiplying the probabilities along the branches. For example, $P(GY) = \frac{5}{10} \cdot \frac{2}{9} = \frac{10}{90}$.

a. Complete the tree diagram for this experiment by listing the outcomes and probabilities. Check your answers with another student.

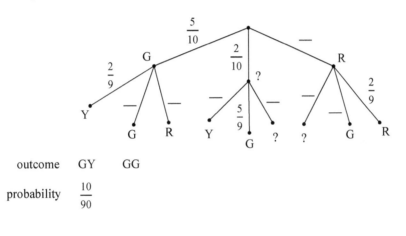

outcome   GY   GG

probability $\dfrac{10}{90}$

b. What is the probability of drawing two green marbles?

c. What is the probability the first marble is yellow and the second marble is red?

d. What is the probability the both marbles have the same color?

e. What is the probability the second marble is yellow, given the first marble is red?

f. What is the probability the second marble is green, given the first marble is green?

g. What is the probability the first marble is yellow and the second marble is green?

3. A container has 12 counters. There are five counters with the number 7, three counters with the number 4, and four counters with the number 2. An experiment consists of drawing two counters, one at a time without replacement, and subtracting the two numbers so that the result is non-negative.

a. Draw the probability tree diagram for the multistage experiment. Check your answers with another student

b. What is the probability the difference is 0?

c. What is the probability the difference is 3 or 5?

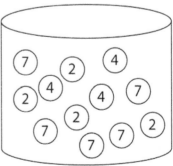

4. An experiment consists of spinning two spinners and adding the numbers together, as follows. First, spin Spinner 1. If the result of Spinner 1 is an even number, then spin Spinner 2. If the result of Spinner 1 is an odd number, then spin Spinner 3.

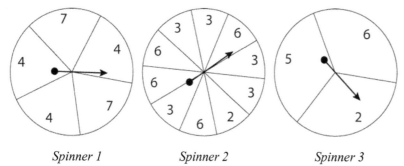

    Spinner 1   Spinner 2   Spinner 3

    a. Draw the tree diagram for this multistage experiment. Check your answers with another student.
    b. What is the probability the outcome is a 10?
    c. What is the probability the outcome is an odd number?

5. Three players are flipping coins to pick a winner. In this game, players A, B, and C each flip a coin. If two players have coins that land heads up and the other player has a coin that lands tails up, then the player with tails up wins. If two players have coins that land tails up and the other player has a coin that lands heads up, then the player with heads up wins. The players repeat the coin toss whenever all three coins have the same outcome.

    a. Suppose the coin for player A has 0.5 probability of landing tails up, the coin for player B has 0.5 probability of landing tails up, and the coin for player C has 0.5 probability of landing tails up. Which player is favored to win the game?

    b. Suppose the coin for player A has 0.3 probability of landing tails up, the coin for player B has 0.65 probability of landing tails up, and the coin for player C has 0.5 probability of landing tails up. Which player is favored to win the game?

    c. Suppose the coin for player A has 0.2 probability of landing tails up, the coin for player B has 0.6 probability of landing tails up, and the coin for player C has 0.75 probability of landing tails up. Which player is favored to win the game?

6. A bag has 6 quarters, 5 dimes, and 4 nickels. An experiment consists of randomly selecting coins, one at a time without replacement, until a quarter is drawn. What is the probability that the third coin is a quarter?

7. Spinner A is spun. If the outcome of Spinner A is an odd number, then Spinner B is spun. If the outcome of Spinner A is an even number, then Spinner C is spun. The results of both spins are added. What is the probability the sum is an even number?

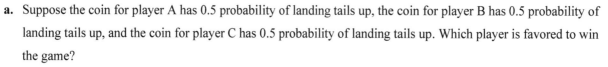

spinner	A		B			C	
outcome	1	2	4	7	2	6	3
probability	0.3	0.7	0.3	0.5	0.2	0.6	0.4

ACTIVITY 9.3.1 Expected Value

Material: sharp pencil and protractor   Name _____

In Section 8.2, we calculated the mean of a collection of numbers using the expression $(x_1 + x_2 + \ldots + x_n)/n$.

Now we calculate the mean when we know the probability associated with each numerical outcome in a sample space. In this situation, the mean is called the "expected value."

1. Consider a spinner with just two outcomes, 3 points and 8 points, as shown.

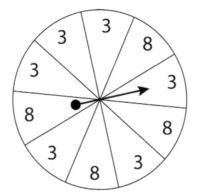

   a. Ideally, in 100 spins we expect to land on 3 exactly _____ times.

   b. Ideally, in 100 spins we expect to land on 8 exactly _____ times.

   c. Ideally, in 100 spins we expect a total of _____ points.

   d. Ideally, we expect the total number of points per spin to be _____ .

   e. Complete the table:

outcome	3	8
probability		

   f. Suppose an experiment has k outcomes $x_1, x_2, .., x_k$ with corresponding probabilities $P(x_1), P(x_2), ..,$ and $P(x_k)$. The expected value of the experiment is the number e defined by $e = x_1 \cdot P(x_1) + x_2 \cdot P(x_2) + \ldots + x_k \cdot P(x_k)$. Calculate the expected value of the spinner.

2. An organization holds a lottery as a fund raiser. They sell 2000 lottery tickets. There are 12 tickets that pay money. One ticket pays $1500, three tickets pay $500 each, and eight tickets pay $100 each.

   a. Calculate the expected value of a lottery ticket.

   b. Suppose the organization sells all of their lottery tickets. How much money altogether would the organization pay out to the winners?

   c. On the average, how much would the organization pay out for each lottery ticket?

   d. Suppose you purchased one ticket. On the average, what is the "expected" payout for the ticket?

   e. Suppose the price of each ticket is $2 and you purchased 15 tickets. How much profit/loss can you expect overall?

3. Calculate the expected value of the spinner shown.

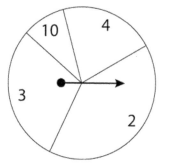

4. The net for a number cube in a board game is shown. The player advances the number of spaces indicated by the number shown on the top face of the number cube. What is the expected value for each roll of the number cube?

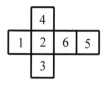

5. The net for a number cube in a board game is shown. The player advances the number of spaces indicated by the number shown on the top face of the number cube. Shaun needs to advance 45 more spaces until she reaches her destination. On the average, how many rolls of the number cube would be required?

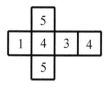

6. The probability that a person in the United States dies in the five-year age interval 70 to 74 years is 0.121699, according to a life table for the total population in a National Vital Statistics Report based National Vital Statistics Reports, June 2010. Suppose Mr. Cane purchases a five-year $200,000 life insurance policy that covers him for the five-year age interval of 70 to 74. Then the insurance company either pays $0 or $200,000.

    a. Use the data to determine the "expected payout" for the policy.

payout	$0	$200,000
probability	0.878301	0.121699

    b. Suppose the insurance company charges $35,000 per life insurance policy.

        How much profit can they expect to make for each policy?

    c. Suppose the insurance company wants to earn an average of $25,000 per life insurance policy.

        How much should they charge for each policy?

7. A game consists of rolling a standard 1-6 number cube. If an odd number is rolled, the player gains that number times $5. But if an even number is rolled, the player loses that number times $4. How much should a player expect to win or lose each roll?

ACTIVITY 9.3.2 Simulations

Material: sharp pencil and table of random digits (p. A12)   Name _____

A **simulation** is a representation or imitation of a process. The experimental results from a simulation provide estimates of theoretical probabilities. Simulations help students develop correct intuitions about probability and predict outcomes. In this activity, you will learn how to use a table of random digits to estimate probabilities. We outline some general steps that you can follow, and they can be modified for other situations.

How to Conduct a Simulation to Approximate the Probability of an Event:

1. Select a model (e.g., a table of random digits).

2. Define an experiment (e.g., randomly selecting a group of 5 digits)

3. Define the event (e.g., the event occurs when 3 of the 5 digits are 0, 1, or 2).

4. Conduct n trials using the model (e.g., randomly picking $n = 30$ groups of 5 digits).

5. Record the frequency f of the number of times the event occurred

   (e.g., mark the groups of digits that indicate the event occurred, then count the number of marks).

6. Compute the quotient f/n as an estimate of the probability of the event.

We will lead you though this process for a situation involving making free throws in basketball.

1. Desiree is a basketball player and has a 65% chance of making a free throw. Use a simulation to estimate the probability that she will make 2 of her next 3 free throws.

   a. Desiree has a 65% chance of making a free throw. There are 100 pairs of digits from 00 to 99. A pair of digits represents a free throw. What could the digits from 00 to 64 represent?

   b. Desiree has a 35% chance of missing a free throw. What could the digits 65 to 99 represent?

   c. Do pairs of digits properly represent the probabilities in the problem?

   d. An experiment consists of selecting 3 pairs of digits to represent the three free throws, such as 34 56 08. How would you determine if the event occurs, that is, if Desiree made exactly 2 of 3 free throws?

   e. Suppose you repeated the experiment, say 30 times ($n = 30$), and obtained the following data from a table of random digits. Circle the groups of digits that represent the occurrence of the event (exactly 2 of 3 free throws made).

10 48 01	50 11 01	53 60 20	11 81 64	79 16 46	69 17 91	41 94 62	59 03 62	07 22 36	84 65 73
25 59 58	53 93 30	99 58 91	98 27 98	25 34 02	93 96 53	40 95 24	13 04 83	60 22 52	79 72 65
76 39 36	48 09 15	17 92 48	30 49 34	03 20 81	42 16 79	30 93 06	24 36 16	80 07 85	61 63 76

   f. Calculate the quotient f/n as an estimate of the probability that Desiree will make 2 of her next 3 free throws.

2. Two players, Huffing and Puffing, miraculously reached the Badminton Championship. It was an odd year, indeed, because Huffing and Puffing, retired elementary school principals, were not known for their speed, footwork, and serves. During the regular season, Huffing won 70% of the games against Puffing. The Championship has a best 3 out of 5 format, which means the first player to win three games wins the championship. Use a table of random digits to estimate the probability that Huffing wins the Badminton Championship.

The steps for conducting a simulation to approximate a probability can be modified for other situations, as demonstrated in the next problem. The key is to make sure that the model represents the probabilities accurately.

3. Mike has a 30% chance of winning a prize in a carnival game. He plays the game until he wins two prizes for his brother and sister. Use a simulation to estimate the number of games he can expect to play to win two prizes by following these steps.

    a. Mike has a 30% chance of winning a prize. There are 10 digits from 0 to 9. Each digit represents a game was played. What should the digits from 0 to 2 represent?

    b. What should the digits 3 to 9 represent?

    c. Do the digits properly represent the probabilities in the problem?

    d. An experiment consists of selecting digits until two of the digits are 0, 1, or 2, such as in the outcome 518941. In the outcome 518941, how many games did Mike have to play to win two prizes?

    e. Repeat the experiment, say 30 times ($n = 30$), and count the number of times the event occurs ($f = ?$). Pick digits from the following table of digits, you may select the digits horizontally.

52162	53916	46369	58586	23216	14513	83149	98736	23495	64350
07056	97628	33787	09998	42698	06691	76988	13602	51851	46104
48663	91245	85828	14346	09172	30168	90229	04734	59193	22178
54164	58492	22421	74103	47070	25306	76468	26384	58151	06646
32639	32363	05597	24200	13363	38005	94342	28728	35806	06912
29334	27001	87637	87308	58731	00256	45834	15398	46557	41135

Enter the 30 simulations in the table below, and write the number of games needed to win two prizes, as illustrated in the first entry.

521 ③									

    f. Average the 30 numbers.

    g. Based on your results, how many games can Mike expect to play to win two prizes?

4. A cereal company puts one prize in every cereal box. There are 4 prizes (say A, B, C, and D), and each prize has the same chance of being in any box. Maria wants two identical prizes for her younger brothers. Use a table of random digits to estimate the number of boxes she should expect to buy to get two identical prizes.

ACTIVITY 10.1.1   Points and Lines

Material: sharp pencil   Name _____

In this activity, we focus on the representations of basic geometric objects using mathematical symbols and their connection to words and diagrams. Fluency with geometry vocabulary is important for defining geometric figures, identifying parts of geometric figures, classifying or grouping geometric figures, and articulating ideas.

	Symbol	**Words**	**Diagram**
1.	A		
2.	$\overrightarrow{AB}$		
3.	$\overline{AB}$		
4.	AB		
5.	ABC		
6.	$\angle B$		
7.	$m\angle B$		
8.	$\overleftrightarrow{AB} \parallel \overleftrightarrow{CD}$		
9.	$\overleftrightarrow{AB} \perp \overleftrightarrow{CD}$		

10. A **postulate** is a property that we accept as true because it is intuitive, leads to interesting consequences, and cannot be proven (or the proof is beyond the scope of the course). There are five postulates in geometry that are useful for geometrical reasoning. Match the postulates to the diagrams.

   a. **Postulate 1.** Given two points, there is exactly one line containing the two points.

   b. **Postulate 2.** Given three noncollinear points, there is exactly one plane containing the three points.

   c. **Postulate 3.** Given two points in a plane, the line containing the two points lies in the plane.

   d. **Postulate 4.** Given two intersecting planes, their intersection is a line.

   e. **Postulate 5.** There is a one-to-one correspondence between the points on a line and the points on the real number line.

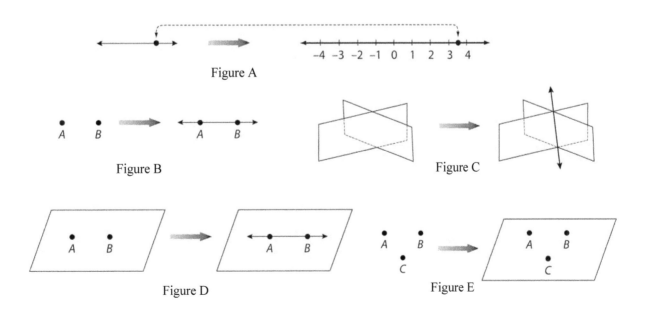

Figure A

Figure B   Figure C

Figure D   Figure E

11. Explain why a line must have infinitely many points.

12. A line has infinitely many points. Suppose a line has 12 points that are labeled as $A_1, A_2, \ldots, A_{12}$.

    a. Write two names for the line using the labeled points.

    b. How many names can you write for the line using the labeled points?

13. A plane has infinitely many points. Suppose a plane has 12 points that are labeled as $A_1, A_2, \ldots, A_{12}$.

    a. Write two names for the plane using the labeled points.

    b. How many names can you write for the plane using the labeled points?

ACTIVITY 10.1.2 Angle Measure

Material: sharp pencil and protractor   Name _____

We can use a ruler to quantify the length of line segment. For example, the length of the line segment shown is $3\frac{1}{2}$ units.

An angle can be formed by two line segments with a common endpoint, two rays with a common endpoint, or two intersecting lines. The "corner" of the angle is called the *vertex*. The following diagram shows three angles: $\angle A$, $\angle B$, and $\angle C$. What attribute of an angle can we quantify? Think about the three angles below. How do they differ? What could be measured?

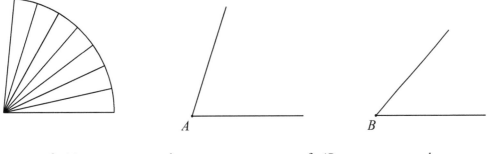

The main difference among the angles shown is how much "openness" the angles have. In this activity, we focus on how to measure the openness of an angle. To quantify the openness of the angle, we need something to compare it to. Let's use a wedge. We shall say that the openness of the angle is the number of wedges that fit in the interior of the angle, side by side, without any gaps or overlap.

1.   Trace the angles on scratch paper and determine the number of wedges that fit in the angle.

measure of $\angle A$ = _____ wedges   measure of $\angle B$ = _____ wedges

2.   List a disadvantage of using a wider wedge, such as the one shown in Figure 1, to measure an angle.

3.   List an advantage of using a thinner wedge, such as the one shown in Figure 2, to measure an angle.

Figure 1   **Figure 2**

One degree (1°) represents a wedge when a circle is subdivided into 360 equal sized wedges (also called sectors). The following diagram shows what such a circle would look like, along with a close-up of a portion of the circle.

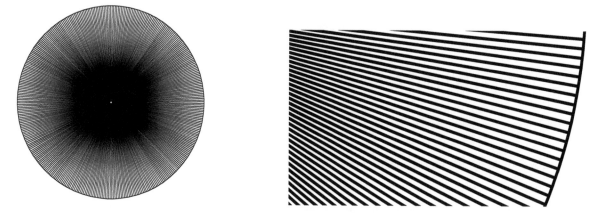

We use a measuring tool called a **protractor** to quantify the *openness* of an angle. Most protractors have the shape of a semicircle and measure angles between 0° and 180°. To measure the openness of an angle, place the **center mark** of the protractor at the vertex of the angle and the **zero line** of the protractor along one side of the angle. Then record the measurement associated with the hash mark indicated by the other side of the angle. Use the symbol ∠ABC for the angle itself, and use the symbol m∠ABC for the measure.

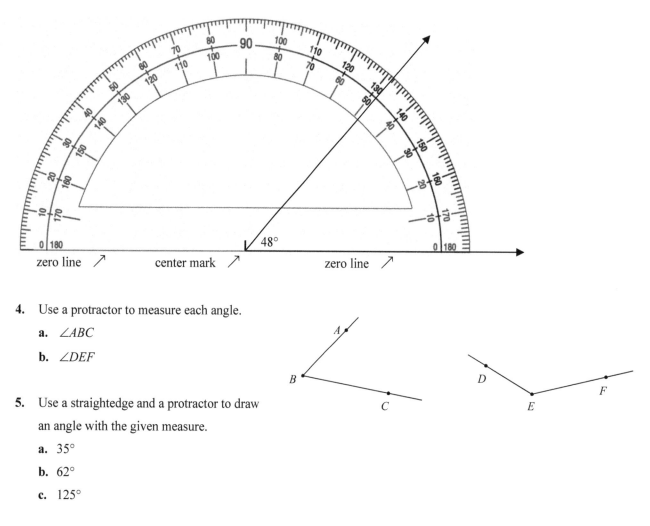

4. Use a protractor to measure each angle.

   a. ∠ABC

   b. ∠DEF

5. Use a straightedge and a protractor to draw an angle with the given measure.

   a. 35°

   b. 62°

   c. 125°

6. The segment addition postulate allows you to determine the length of a line segment indirectly using subtraction. Use the segment addition postulate to find the unknown length.

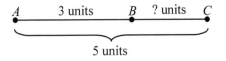

7. The angle addition postulate allows you to determine the measure of an angle indirectly using subtraction. Use the angle addition postulate to find the unknown angle measurement.

   a. m∠DBC =   **b.** m∠GFH =

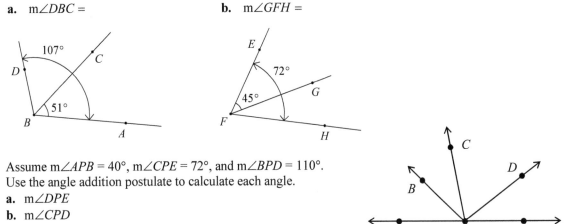

8. Assume m∠APB = 40°, m∠CPE = 72°, and m∠BPD = 110°. Use the angle addition postulate to calculate each angle.
   a. m∠DPE
   b. m∠CPD
   c. m∠BPC

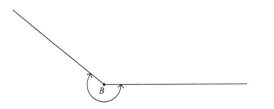

9. Draw an example of each angle, and then use a protractor to measure the angle.

   a. acute angle

   b. obtuse angle

10. ∠B is an example of a reflex angle.

   a. How would you define a reflex angle?

   b. Discuss how you could use a protractor (in the shape of a semi-circle) to measure a reflex angle.

   c. Use a protractor to measure the reflex angle ∠B

11. Use the Solve an Equation strategy to solve the problem. The sum of two angles is 180°.

   The larger angle is 10° more than four times the smaller angle. Find the measure of each angle.

12. Use the Solve an Equation strategy to solve the problem. The sum of two angles is 216°.

   The larger angle is 8° less than three times the smaller angle. Find the measure of each angle.

ACTIVITY 10.2.1   Angles and Paper Folding

Material: sharp pencil, scratch paper, and protractor   Name _____

Geometry is a branch of mathematics that deals with figures and their size and shape. You can use paper folding to explore two-dimensional geometry. This activity will investigate properties of angles when created by intersecting lines using the method of folding paper. Each member of the group will use paper folding constructions and then compare their findings with the group. Use ordinary sheets of paper and make all folds as straight as possible. Crease each fold before opening the sheet. To make the pictures more defined, trace each fold with a straightedge and pencil.

1. Fold one sheet of paper. Open the paper and describe what is geometrically created by this crease. The piece of paper has one line folded on it. Fold one more line so that the two lines intersect and create the image of an × as shown in Figure 1. The intersection of two lines creates four angles. After tracing the lines with a pencil and straightedge, label the four angles with the numbers 1, 2, 3, and 4 in a clockwise order. Then measure each of the four angles and state their measurements below.

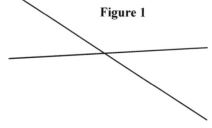

Figure 1

2. Write your measurements for angles 1, 2, 3, 4:
    $m\angle 1 =$

    $m\angle 2 =$

    $m\angle 3 =$

    $m\angle 4 =$

3. Describe any patterns in the measurements of these angles.

4. Compare the patterns in your measurements of the angles to those found by your group members. What patterns seem to hold?

5. Adjacent angles are two angles that share a common side. List the four pairs of adjacent angles from your paper folding.

6. Supplementary angles are two angles that have a sum of measures equal to 180°. List the four pairs of supplementary angles from your paper folding.

7. Vertical angles are two nonadjacent angles formed from two intersecting lines. List the two pairs of vertical angles from your paper folding.

8. What can be said about all pairs of vertical angles?

ACTIVITY 10.2.2   Two Lines Cut by a Transversal and Angle Relationships

Material: sharp pencil, protractor, and straightedge   Name _____

A **transversal** is a line that cuts (intersects) two lines at different points. The goal of this activity is learn how angles created by a transversal are related.

1. Figure 1 shows two parallel lines cut by transversal t.

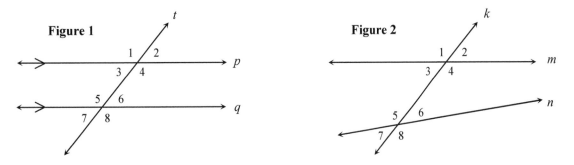

   a. Choose a pair of corresponding angles in Figure 1. Use a protractor to measure the two angles that you chose.

   b. Compare your results with two other students in the class.

   c. Use logical reasoning (instead of a protractor) to determine the measures of all the other labeled angles.

   d. How are two corresponding angles related?

   e. Complete the conjecture that relates parallel lines and corresponding angles:

       If two lines cut by a transversal are parallel, then

       _____

2. Figure 2 shows two nonparallel lines cut by transversal k.

   a. Choose a pair of corresponding angles in Figure 2. Use a protractor to measure the two angles that you chose.

   b. Compare your results with two other students in the class.

   c. Use logical reasoning (instead of a protractor) to determine the measures of all the other labeled angles.

   d. How are two corresponding angles related?

   e. Complete the conjecture that relates nonparallel lines and corresponding angles:

       If two lines cut by a transversal are nonparallel, then

       _____

3. What can you say about $\angle 1$ in Figure 3? Justify your answer.

4. What can you say about $\angle 2$ in Figure 4? Justify your answer.

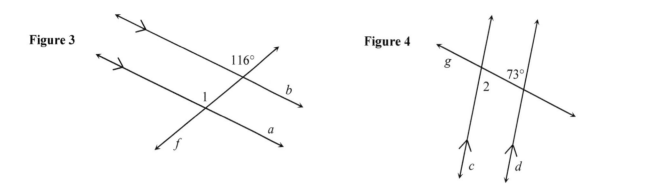

Figure 3

116°

1

b

f

a

Figure 4

g

73°

2

c   d

5. Figure 5 shows two lines cut by transversal *t* along with the measurements of a
   pair of corresponding angles.

   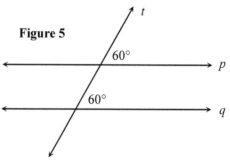

   Figure 5

   60°

   t

   p

   60°

   q

   a. Use logical reasoning (instead of a protractor) to determine the measures of
      the other angles in Figure 5.

   b. How are two corresponding angles related?

   c. Do lines *p* and *q* appear parallel or nonparallel?

6. Complete the conjecture that relates parallel lines and corresponding angles:

   If two lines cut by a transversal have a pair of corresponding angles that are congruent, then

   _____

7. Figure 6 shows two lines cut by transversal *k* along with the measurements of a
   pair of corresponding angles.

   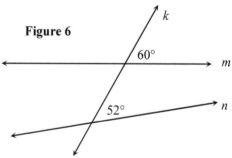

   Figure 6

   k

   60°

   m

   52°

   n

   a. Use logical reasoning (instead of a protractor) to determine the measures of
      the other angles in Figure 6.

   b. How are two corresponding angles related?

   c. Do lines *m* and *n* appear parallel or nonparallel?

8. Complete the conjecture that relates nonparallel lines and corresponding angles:

   If two lines cut by a transversal have a pair of corresponding angles that are not congruent, then

   _____

9. Sketch two parallel lines without using the arrow marks to signify parallelism.

10. Maria said lines *u* and *v* are nonparallel lines. Use a straightedge and protractor to support her claim.

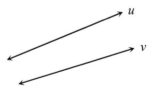

u

v

ACTIVITY 10.2.3   Types of Quadrilaterals

Material: sharp pencil   Name _____

In this activity, we focus on the representations of basic geometric objects using mathematical symbols or diagrams.

1. A **polygon** is simple, closed curved curve that consists of line segments. A convex polygon does not have any indentations. Draw a convex polygon. Identify the following parts related to a polygon.

    a. polygon

    b. interior of the polygon

    c. exterior of the polygon

    d. interior angle

    e. side

    f. diagonal

    g. vertex

    h. exterior angle

2. A **quadrilateral** is a polygon with four sides. Draw a convex quadrilateral.

    a. How many interior angles does it have?

    b. How many exterior angles does it have?

    c. How many diagonals does it have?

3. A **pentagon** is a polygon with five sides. Draw a convex pentagon.

    a. How many interior angles does it have?

    b. How many exterior angles does it have?

    c. How many diagonals does it have?

4. A **hexagon** is a polygon with six sides. Draw a convex hexagon.

    a. How many interior angles does it have?

    b. How many exterior angles does it have?

    c. How many diagonals does it have?

5. A **heptagon** is a polygon with seven sides. Draw a convex heptagon.

    a. How many interior angles does it have?

    b. How many exterior angles does it have?

    c. How many diagonals does it have?

Write a definition for each type of quadrilateral in the following problems. The markings on the quadrilateral indicate properties that your definition must support. Then show your definition to another student. If the student can draw a polygon that abides by your definition but doesn't represent the polygon that you defined, then your definition needs revision. Or, if the student can make the definition more concise, then the definition needs revision. This activity will help you appreciate the need for concise and precise definitions.

6.   A **parallelogram** is a convex quadrilateral …

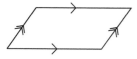

7.   A **rectangle** is …

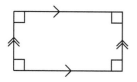

8.   A **square** is …

9.   A **rhombus** is a convex quadrilateral …

10. A **kite** is a convex quadrilateral …

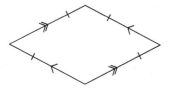

11. A **trapezoid** is a convex quadrilateral …

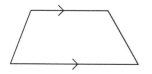

ACTIVITY 10.2.4   Sum of the Measures of the Interior Angles of a Polygon

Material: sharp pencil   Name _____

In this activity, we focus on the sum of the measures of the interior angles of a polygon.

Figure 1 shows the interior angles of a triangle. What is the sum of the measures of the interior angles? The goal is to find $m\angle 1 + m\angle 2 + m\angle 3$.

Figure 1

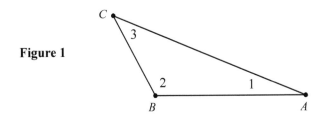

1. Follow these steps to find the sum of the angles of the triangle shown.

    a. Pick a side of the triangle, and draw a line through both endpoints of that line segment.

    b. Locate the vertex of the triangle that does not belong to the side that you picked in part (a). Then draw a line through that vertex such that the line is parallel to the side you picked.

    c. Use the fact that "if a transversal cuts parallel lines, then any pair of alternate interior angles are congruent" to mark some unmarked angles with 1, 2, or 3 as appropriate.

    d. Use the diagram to write an equation.

2. For each polygon in Figure 2, draw all possible diagonals from vertex A to partition the each polygon into triangles. Use the diagram to find the sum of the measures of the interior angles of the polygon.

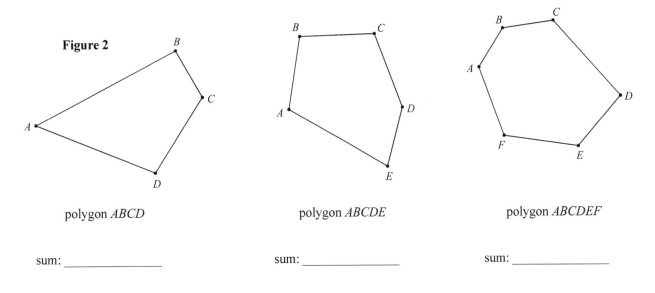

Figure 2

polygon $ABCD$ polygon $ABCDE$ polygon $ABCDEF$

sum: _____   sum: _____   sum: _____

3.   Write the angle sums in the table. Look for a pattern to find the sum of the interior angles of an *n*-gon, which is a polygon with *n* sides.

types of polygon	number of angles	sum of the interior angles
triangle	3	180°
quadrilateral	4	
pentagon	5	
hexagon	6	
heptagon	7	
octagon	8	
nonagon	9	
decagon	10	
n-gon	*n*	

4.   Determine the sum of the measures of the interior angles of each polygon.

   a. A polygon with 30 sides.

   b. A polygon with 40 sides.

   c. A polygon with 50 sides.

5.   A **regular polygon** is a polygon such that all the sides have the same length and all the interior angles have the same measure.

   a. Determine the measure of each interior angle of a regular polygon with 5 sides.

   b. Determine the measure of each interior angle of a regular polygon with 6 sides.

   c. Determine the measure of each interior angle of a regular polygon with 7 sides.

   d. What happens to the measure of an interior angle of a regular polygon as the number of sides increases?

6.   Do the following.

   a. Determine the measure of each interior angle of a regular polygon with 20 sides.

   b. Determine the measure of each interior angle of a regular polygon with 60 sides.

   c. Determine the measure of each interior angle of a regular polygon with 100 sides.

7.   Do the following.

   a. Determine the measure of each exterior angle of a regular polygon with 5 sides.

   b. Determine the measure of each exterior angle of a regular polygon with 6 sides.

   c. Determine the measure of each exterior angle of a regular polygon with 7 sides.

ACTIVITY 10.3.1 Relationships in Prisms and Pyramids

Material: sharp pencil, scissors, tape, nets (pp. A13 and A14)   Name _____

The purpose of this activity is to learn relationships among the edges, faces, and vertices of prisms and pyramids.

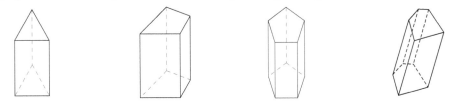

regular triangular prism   quadrilateral prism   regular pentagonal prism   pentagonal prism

1. Use scissors to cut the net on page A13 and form the regular pentagonal prism.

2. Identify the part of the prism indicated by the dot(s), shading, or boldface. Use all vocabulary words that apply:
    apex, base, edge, face, lateral face, vertex, vertices.

    a.   b.   c.   d.   e.   f.

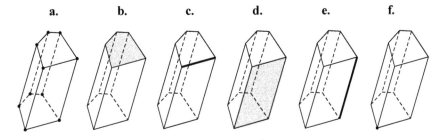

3. Let n represent the number of sides of a polygonal base. Let V represent the number of vertices of the prism.
    Let F represent the number of faces (bases and lateral faces) of the prism. Let E represent the number of edges of the
    prism. Use the prisms shown to complete the table.

n	V	E	F
3			
4			
5			

4. Based on the pattern in the table, write an equation involving V and n: $V =$

5. Is there a prism with 846 vertices? Explain.

6. Is there a prism with 651 vertices? Explain.

7. Based on the pattern in the table, write an equation involving F and n: $F =$

8. Is there a prism with 521 faces? Explain.

9. Is there a prism with 638 faces? Explain.

10. Based on the pattern in the table, write an equation involving E and n: $E =$

11. Is there a prism with 141 edges? Explain.

12. Is there a prism with 257 edges? Explain.

13. Suppose a prism has 261 edges. How many vertices does it have? Explain.

14. Suppose a prism has 472 vertices. How many faces does it have? Explain.

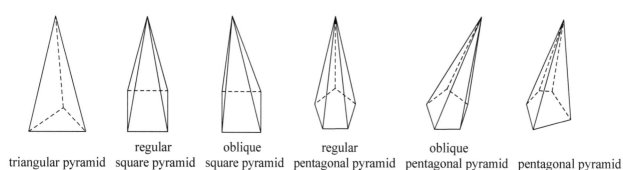

triangular pyramid   regular square pyramid   oblique square pyramid   regular pentagonal pyramid   oblique pentagonal pyramid   pentagonal pyramid

15. Use scissors to cut the net on page A14 to make the regular pentagonal pyramid.

16. Identify the part of the pyramid indicated by the dot(s), shading, or boldface. Use all vocabulary words that apply: apex, base, edge, face, lateral face, vertex, vertices.

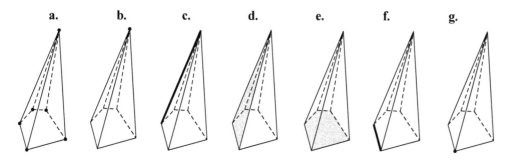

17. Let n represent the number of sides of the polygonal base. Let F represent the number of faces (bases and lateral faces) of the pyramid. Let V represent the number of vertices of the pyramid. Let E represent the number of edges of the pyramid. Use the pyramids shown to complete the table.

n	V	E	F
3			
4			
5			

18. Based on the pattern in the table, write an equation involving V and n: $V =$

19. Is there a pyramid with 301 vertices? Explain.

20. Is there a pyramid with 724 vertices? Explain.

21. Based on the pattern in the table, write an equation involving F and n: $F =$

22. Is there a pyramid with 200 faces? Explain.

23. Is there a pyramid with 601 faces? Explain.

24. Based on the pattern in the table, write an equation involving E and n: $E =$

25. Is there a pyramid with 612 edges? Explain.

26. Is there a pyramid with 387 edges? Explain.

27. Suppose a pyramid has 456 edges. How many vertices does it have? Explain.

28. Suppose a pyramid has 311 faces. How many edges does it have? Explain.

ACTIVITY 11.1.1   Basics of Measurement

Material: sharp pencil   Name _____

Measurement is a quantitative comparison of an attribute of an object to the chosen measurement unit. For example, the length of the line segment shown is 5 paper clips. The attribute of the line segment is length and the measurement unit is the paper clip. In this activity, we focus on the basics of measurement.

1. Number the steps (1, 2, 3, or 4) in the process of measurement typically used in the elementary classroom in the correct order so that 1 is the first step, 2 is the second step, and so on.

    _____ Choose an appropriate measurement unit.

    _____ Report the measurement with a numerical value and a unit of measurement.

    _____ Choose an attribute of an object to measure.

    _____ Place the measurement unit end to end to cover the attribute without gaps or overlaps, and then count the number of repetitions.

2. Is the length of the pencil 4 papers clips? Why or why not?

    a.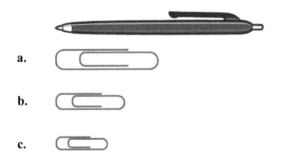

    b.

    c.

3. Measure the length of the pen using the measurement unit shown.

    a.

    b.

    c.

4. Use your results in Problem 3 to discuss the relationship between the size of the unit of measurement and the number of units needed to measure the length of the pen.

5. A **measurement system** is a collection of measurement units and rules for relating the various units. A table is useful way of displaying the units. The table shows a fictitious measurement system. As a group, use *dimensional analysis* to convert the measurements to the indicated unit. Round the final answer to the nearest tenth.

Measurement unit	Relationship
clip	2 clips = 3 pens
pen	5 pens = 9 zaps
zap	

a. 7 clips = _____ pens

b. 4 zaps = _____ pens

c. 15 clips = _____ zaps

6. Use dimensional analysis to convert the measurements. Round the final answer to the nearest 0.1.

Measurement unit	Relationship
clip	7 clips = 4 pens
pen	
zap	5 zaps = 3 clips

a. 10 clips = _____ zaps

b. 20 pens = _____ clips

c. 15 pens = _____ zaps

d. 14 zaps = _____ clips

7. Measure the length of the pencil using each of the three rulers, rounding to the nearest mark on the ruler.

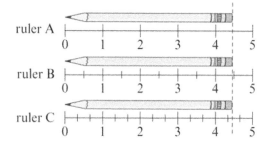

8. Measurements are typically inexact due to rounding.

a. Which measurement in Problem 7 is the most accurate? Explain.

b. If you were a carpenter and had to choose a ruler from the three rulers in Problem 7, which one would you choose? Explain.

9. Some students think the length of the pencil is $12\frac{1}{2}$ units.

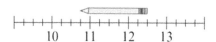

a. How did they arrive at the answer?

b. How would you correct their thinking?

10. Determine the smallest unit of measurement (clip, pen, or zap) in the measurement system in Problem 6.

ACTIVITY 11.2.1   Perimeter and π

Material: sharp pencil, string, metric ruler, and   Name _____

circular objects such as lids or cans

The **perimeter** of a simple closed curve, such as a circle or polygon, is the length of the curve. The purpose of this activity is to use the idea of perimeter to help explain what π means. The **circumference** of a circle is the conventional name for the perimeter of the circle. The word circumference may refer to the curve itself or the measure of the curve. This is similar to the dual nature of the word diameter, which refers to either a line segment or measurement.

For this activity, gather several circular objects, such as a circular lid to a tennis can, a circular lid to a cylindrical container of oatmeal, a circular paper plate, or a circular lid of a plastic food container. Wrap string around the circular object along the circumference, and then measure the length of the string required. Record your result in the table in the column for C. Extend the string across the circular object in a straight line through the center of the circle, and then measure the length of the string required. Record your result in the table in the column for d. Then use a calculator to calculate the quotient C/d. Record your result in the table in the column for C/d, rounding to the nearest tenth.

Actual measurements of some circular objects

object	circumference C	diameter d	quotient $\dfrac{C}{d}$
can	210 mm	66 mm	3.2

1.   What did you notice about the values of the quotients $\frac{C}{d}$?

2.   Ideally, the quotients would all be constant (identical) for the various circles. The symbol π denotes the constant value of the quotient $\frac{C}{d}$, that is, $\pi = \frac{C}{d}$

   a.   Solve the equation for C.

   b.   Express C in terms of the radius r of the circle.

3. π has infinitely many digits after the decimal point without any repeating pattern; here are a few leading digits:

   $\pi = 3.14159265358979323846 \dots$. Use a calculator key for π or use $\pi \approx 3.1416$ in your calculations.

   a.   Determine the circumference of a circle with radius 20.8 cm. Round to the nearest tenth.

   b.   A circle has circumference 50π cm. Determine the radius of the circle. Round to the nearest tenth.

   c.   A circle has circumference 38.4 cm. Determine the radius of the circle. Round to the nearest tenth.

4.   Determine the circumference of the circle. You can use a protractor or ruler as needed.

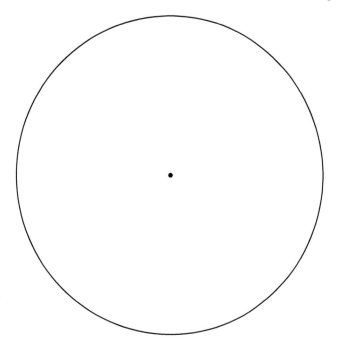

5.   A dog chased a cat around a circular garden with radius 25 feet. The cat ran a total of 1445 feet.
      How many full laps around the garden did the cat run?

6.   A nickel is rolled around the perimeter of a rectangle that has length 270 mm and width 110 mm. The diameter of a
      nickel is 21.21 mm. How many full revolutions will the nickel make until it returns to its starting position?

start

ACTIVITY 11.2.2   Area of Common Shapes

Material: sharp pencil and metric ruler   Name _____

The purpose of this activity is to use a dot grid or ruler to obtain measurements and then apply the formulas for area.

1.   Use the formula for the area of a triangle to find the area of △ *ABC*.

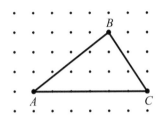

2.   Use the formula for the area of a triangle to find the area of △ *DEF*.

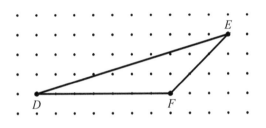

3.   Use the formula for the area of a parallelogram to find the area of parallelogram *ABCD*.

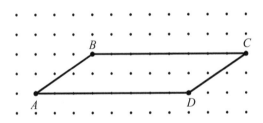

4.   Use the formula for the area of a trapezoid to find the area of trapezoid *EFGH*.

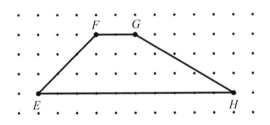

5.   △ *ABC* is shown.

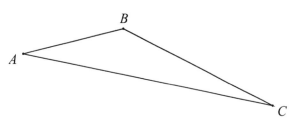

a. Suppose the base is $\overline{AC}$. Use a straightedge to draw the altitude for base $\overline{AC}$.

b. Use a ruler to measure the base and altitude from part (a) and then apply the formula for the area of a triangle to find the area of △ *ABC*. Round to the nearest mm^2 .

c. Suppose the base is $\overline{AB}$. Use a straightedge to draw the altitude for base $\overline{AB}$.

d. Use a ruler to measure the base and altitude from part (c) and then apply the formula for the area of a triangle to find the area of △ *ABC*. Round to the nearest mm^2 .

e. Suppose the base is $\overline{BC}$. Use a straightedge to draw the altitude for base $\overline{BC}$.

f. Use a ruler to measure the base and altitude from part (e) and then apply the formula for the area of a triangle to find the area of △ *ABC*. Round to the nearest mm^2 .

g. Should the answers to parts (b), (d), and (f) be the same? If so, then explain any discrepancies in your answers.

6.   Use the dot grid to draw a triangle with area 6 square units such that the triangle is the given type.

a. acute

b. right

c. obtuse

7.   Discuss how you could calculate the area of the irregular heptagon shown. Then use a ruler and follow your strategy to calculate the area of the heptagon. Round your final answer to the nearest square millimeter.

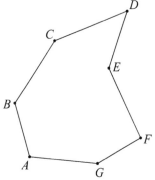

ACTIVITY 11.2.3 Dimensional Analysis and Area

Material: sharp pencil

Name _____

The purpose of this activity is to gain practice using dimensional analysis to convert area measurements.

1. Do the following.

   a. Fill in the blank: 1 yard = _____ feet

   b. The diagram represents 1 square yard, which is a square such that each side has length 1 yard.

   Explain why 1 square yard equals 9 square feet.

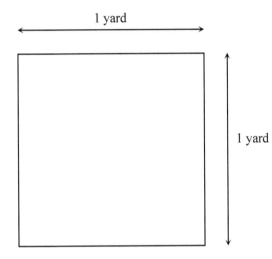

2. **Dimensional analysis** is the process of converting measurements by treating measurements as variables.

   Here's how we would use dimensional analysis to convert 25 yd^2 to square feet (ft^2):

   $$25 \text{ yd}^2 = 25 \text{ yd}^2 \times \left(\frac{3 \text{ ft}}{1 \text{ yd}}\right)^2 = 25 \text{ yd}^2 \times \frac{3^2 \text{ ft}^2}{1^2 \text{ yd}^2} = 25 \text{ yd}^2 \times \frac{9 \text{ ft}^2}{1 \text{ yd}^2} = 25 \cdot 9 \text{ ft}^2 = 225 \text{ ft}^2.$$

   Use dimensional analysis to convert the measurements. Round the final answer to the nearest 0.1.

Measurement unit	Relationship
clip	7 clips = 4 pens
pen	
zap	5 zaps = 3 clips

   a. 12 square clips = _____ square zaps

   b. 18 square pens = _____ square clips

   c. 14 square pens = _____ square zaps

   d. 22 square zaps = _____ square clips

   e. 345 square zaps = _____ square pens

   f. 20 square clips = _____ square pens

ACTIVITY 11.2.4   Motivating the Formula for the Area of a Circle

Material: sharp pencil, scissors, clear tape, circle cutout (p. A15)   Name _____

The purpose of this activity is to motivate the formula for the area of a circle.

1. Figure 1 can be found on page A15.

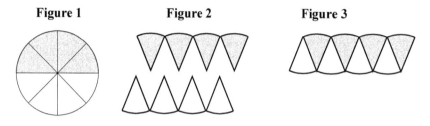

    Figure 1   **Figure 2**   **Figure 3**

    a. Figure 1 shows a circle with radius r divided into eight equal sized sectors. Cut the semicircles into two chains of
    sectors as shown in Figure 2.

    b. Assemble the two chains of sectors to form a bumpy shape as shown in Figure 3. Use tape to secure the two
    chains together. What does this shape look like?

    c. Write a formula that approximates the area of the bumpy shape in Figure 3.

    d. How are the areas of the circle in Figure 1 and the bumpy shape in Figure 3 related?

2. The diagram shows a circle with radius r units and a square with sides of length
    r units. Such a square is called a "radius square."

    a. Trace several radius squares on some scratch paper and then cut and tear the
    squares and place the pieces in the circle without gaps or overlaps, to the best of
    your ability. Approximately how many radius squares will fit in the circle,
    without gaps or overlaps?

    b. Exactly how many radius squares will fit in the circle, without gaps or overlaps?

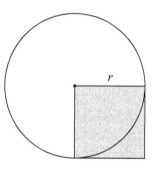

3. Use a ruler and the formula for the area of a circle to find the area of the circle with center A.

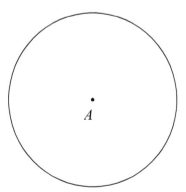

4. The ancient Egyptians approximated the area of a circle using an octagon. Figure 4 shows a circle with diameter d units. Figure 5 shows a square with sides of lengths d units. Figure 6 shows an octagon created by removing the four isosceles triangles in the corners.

Figure 4   **Figure 5**   **Figure 6**

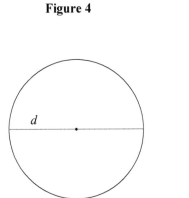 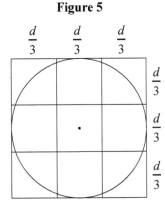 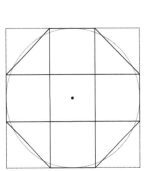

a. What is the area of the square that surrounds the circle?

b. What is the area of each isosceles triangle that was removed?

c. Use your answers in parts (a) and (b) to express the area of the octagon using the variable d.

d. Express the area of the octagon using the variable r, which is the radius of the circle.

e. Compare the areas of the circle and the octagon for various values of r by completing the table. Round the final answers to the nearest 0.01.

radius (units)	A, area of circle (square units)	O, area of octagon (square units)	relative error $\left\lvert \dfrac{A-O}{A} \right\rvert \times 100\%$
5	78.54	77.78	0.97%
12			
85			
142			
270			

ACTIVITY 11.3.1 Pythagorean Triples

Material: sharp pencil   Name _____

Three whole numbers a, b, and c that satisfy the equation $a^2 + b^2 = c^2$ are called a "Pythagorean triple."
The numbers are the measurements of the sides of a right triangle with legs having lengths a and b units
and hypotenuse c units. This activity focuses on three ways to generate a Pythagorean triple.

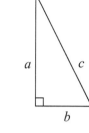

1. Do the following.

    a. Verify that the numbers 3, 4, and 5 form a Pythagorean triple.

    b. If we multiply 3, 4, and 5 each by 7, then we get 21, 28, and 35.

    Verify that 21, 28, and 35 form a Pythagorean triple.

    c. Use multiplication to find another Pythagorean triple.

2. The Greeks gave the following formula for generating a Pythagorean triple: $v^2 - u^2$, $2uv$, $v^2 + u^2$, with $v > u$.

    a. Suppose $v = 8$ and $u = 5$. Use the formula to find the three numbers.

    b. Verify that the three numbers in part (a) form a Pythagorean triple.

    c. Use the formula to generate your own Pythagorean triple.

    d. Verify that the three numbers in part (c) form a Pythagorean triple.

3. A teacher described the following strategy for finding a Pythagorean triple
    (see Sullins, 2004 in the textbook bibliography).

    First, pick an odd number: 11

    Second, square it: 121

    Third, determine two consecutive numbers whose sum is the squared number: 60, 61

    Then 11, 60, 61 is a Pythagorean triple: $11^2 = 121$, $60^2 = 3600$, $61^2 = 3721$; $11^2 + 60^2 = 121 + 3600 = 3721 = 61^2$,

    a. Suppose you choose the odd number 5. Follow these steps to find the other two values in the Pythagorean triple.

    b. Suppose you choose the odd number 9. Follow these steps to find the other two values in the Pythagorean triple.

    c. Follow these steps with your own odd number.

    d. Suppose you know this method produces the Pythagorean triple n, 112, and 113. What is n?

    e. Suppose you know this method produces the Pythagorean triple x, y, and 2813, where 2813 is the largest number
    of the three numbers. What are x and y?

ACTIVITY 11.3.2 The Triangle Inequality

Material: sharp pencil, scissors, line segment arrays (p. A16) Name _____

Suppose you are given three line segments with lengths 4 units, 7 units, and 9 units. Can you form a triangle with these line segments? This activity explores conditions that help you decide if you can form a triangle with three given lengths.

Figure 1 shows three line segments with lengths 4 units, 7 units, and 9 units. Figure 2 shows that we can indeed form a triangle with sides having lengths 4 units, 7 units, and 9 units. Figure 3 compares the longest length to the sum of the two shorter lengths. We see that the longest side is shorter than the sum of the two shorter lengths.

Figure 1. Line segments of lengths 4 units, 7 units, and 9 units.

Figure 2. Forming a triangle with the three line segments.

Figure 3. Comparing the longest length to the sum of the two shorter lengths.

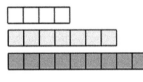

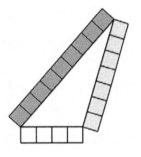

1. In each of the following problems, use scissors to cut line segments of the given lengths from the line segment arrays on page A16. Use the darker color for the longest side, use the lighter color for the second longest side, and use the unshaded color for the shortest side, as demonstrated in Figure 1. Your goal is to try to create a triangle from the line segments, as demonstrated in Figure 2. Be sure to connect the corners of the arrays as shown in Figure 2. Then compare the longest length to the sum of the two shorter lengths, as demonstrated in Figure 3. Complete the table.

lengths	Were you able to form a triangle with the given lengths?	Was the longest line segment shorter than the sum of the two shorter lines segments?
a. 4 units, 7 units, and 10 units	Yes	Yes
b. 5 units, 6 units, and 9 units		
c. 3 units, 4 units, and 5 units		
d. 4 units, 5 units, and 6 units		
e. 4 units, 3 units, and 10 units		
f. 3 units, 6 units, and 13 units		
g. 10 units, 5 units, and 5 units		

2. In some cases, you were able to form a triangle with the given line segments. What did they have in common?

3. In some cases, you were unable to form a triangle with the given line segments. What did they have in common?

4. Suppose you are given three line segments with lengths a units, b units, and c units, where the longest side is c units. How would determine whether the line segments can be used to form a triangle?

5. Can three line segments with lengths 14 feet, 25 feet, and 36 feet be used to form a triangle? Explain.

6. Can three line segments with lengths 63 feet, 140 feet, and 28 feet be used to form a triangle? Explain.

7. Work in groups of three students for this problem. The first student announces the length of a line segment. Then the second student announces the length of another line segment. Then the third student must announce the length of another line segment such that all three line segments can be used to form a triangle. As a group, decide if the third person in the group gave a correct answer. Then brainstorm other possible answers that the third student could have given. Take turns in giving the measurements of the first, second, and third line segment.

If you know that the longest line segment has length c units, then you would just need check the inequality $c < a + b$ to determine if three line segments that have lengths a, b, and c units can be used to form a triangle. For example, can three line segments with lengths 20 feet, 32 feet, and 15 feet be used to form a triangle? $32 < 20 + 15$, so you can form a triangle with line segments with lengths 20 feet, 32 feet, and 15 feet. This is a special case of a more general result called the **Triangle Inequality**: "Line segments with lengths a, b, and c units can be used to form a triangle, if and only if, all three inequalities hold $a < b + c$, $b < a + c$, and $c < a + b$."

Suppose a triangle has sides of length 20 cm and 25 cm. What are possible lengths of the third side of the triangle? Let the third side have length n units. According to the Triangle Inequality, line segments with lengths 20 cm, 25 cm, and n cm can be used to form a triangle, if and only if, the following three inequalities hold: $20 < 25 + n$, $25 < 20 + n$, and $n < 20 + 25$. Then $-5 < n$, $5 < n$, and $n < 45$. Then $5 < n < 45$. The third side must have a length between 5 cm and 45 cm. Follow this example to solve the next three problems.

8. A triangle has sides of length 40 cm and 30 cm. What are possible lengths of the third side of the triangle? (**Hint**: Let the third side have length n units. Then the following inequalities must hold according to the Triangle Inequality: $n < 40 + 30$, $30 < n + 40$, and $40 < n + 30$. Then simplify these inequalities to learn more about n).

9. A triangle has sides of length 22 cm and 46 cm. What are possible lengths of the third side of the triangle?

10. A triangle has sides of length 18 cm and 32 cm. What are possible lengths of the third side of the triangle?

ACTIVITY 11.3.3   Area of a Regular Polygon

Material: sharp pencil and metric ruler   Name _____

The purpose of this activity is to derive and apply the formula for a regular polygon. Students will use a metric ruler to obtain the measurements needed in the formula.

1. Figure 1 illustrates three line segments (side, apothem, and radius) associated with every regular polygon. Fill in the blanks.

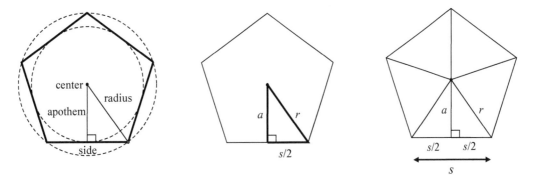

   a. A(n) _____ is a line segment of the polygon.

   b. The _____ is a line segment that extends from the center of the polygon to the midpoint of a side (the radius of the inner circle).

   c. The _____ is a line segment that extends from the center of the polygon to a vertex of the polygon (the radius of the outer circle).

2. Think of the following three line segments associated with a regular polygon: side, apothem, and radius.

   a. Which line segment bisects a side of the regular polygon?

   b. Which line segment is perpendicular to a side of the regular polygon?

   c. Which line segment bisects an interior angle of the regular polygon?

3. Write an equation that shows how the variables a, s, and r are related.

4. The diagram shows a regular pentagon (with the center).

   a. Divide the pentagon into five congruent triangles.

   b. Tell what information you need to know to find the area of one triangle.

   c. Write a formula for the area B of a regular pentagon in the form $B = ?$.

   d. Generalize your formula for the area of a regular n-gon. Write the formula in the form $B = ?$, where the right-hand side of the formula contains just two variables.

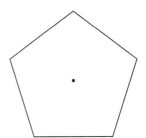

5. A regular pentagon has sides with lengths 14 cm and apothem 9.6 cm. Find the area of the regular pentagon.

6. A regular pentagon has a perimeter of 140 cm and radius 23.8 cm. Find the area of the regular pentagon.

7. A regular octagon has perimeter 240 cm and radius 39.2. Find the area of the regular octagon.

8. Use a metric ruler to find the area of the regular pentagon using your measurements for

    a. a and s.

    b. r and s.

    c. a and r.

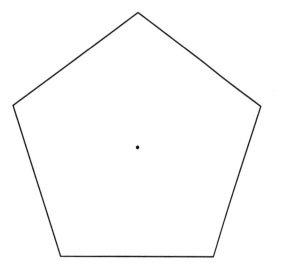

9. Use a metric ruler to find the area of the regular octagon using your measurements for

    a. a and s.

    b. r and s.

    c. a and r.

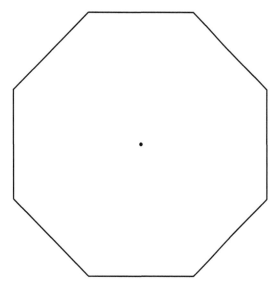

ACTIVITY 11.4.1   Relating the Volume of a Pyramid and Prism

Material: sharp pencil, scissors, tape, construction paper,   Name _____

rice, and nets (pp. A17 and A18)

The purpose of this activity is to relate the volume of a pyramid to the volume of a prism to develop a formula for the volume of a pyramid. We assume that you already know the formula for the volume of a prism.

The two nets for this activity are on pages A17 and A18. Place a piece of construction paper under the net for the pyramid. Use a pencil or pen and trace the net of the pyramid. Be sure to press on the pencil or pen so that it leaves indented lines on the construction paper. Then use a pair of scissors to cut out the net for the pyramid on the construction paper. Remove the shaded region within the base and cut along the dashed lines of the base to form flaps that can be folded inward. Secure the faces and flaps with tape. Repeat this procedure for the net of the prism using another sheet of construction paper. Note: the nets can be formed without the use of construction paper.

1. Compare the height of the pyramid to the height of the prism. What did you notice?

2. Compare the base of the pyramid to the base of the prism. What did you notice?

3. Fill the pyramid with rice. Make sure the rice is leveled at the top. Then pour the rice in the prism. Repeat this procedure until the prism is filled with rice so that the rice is leveled at the top. Keep track of the number of times you poured the rice into the prism.

4. How many scoops of rice from the pyramid did you pour into the prism?

5. Suppose a pyramid has volume 40 cubic units. What is the volume of a prism that has the same base and same height of the pyramid?

6. Suppose a pyramid has volume 300 cubic units. What is the volume of a prism that has the same base and same height of the pyramid?

7. Suppose a prism has volume 60 cubic units. What is the volume of a pyramid that has the same base and same height of the prism?

8. Suppose a prism has volume 150 cubic units. What is the volume of a pyramid that has the same base and same height as the prism?

9. Suppose a prism has volume 210 cubic units. What is the volume of a pyramid that has the same base and same height as the prism?

10. Suppose the area of a base of a prism is B square units and the prism has height h units.
    a. Write an equation for the volume V of the prism.
    b. What is the volume V of a pyramid that has the same base and same height as the prism?

11. Suppose a pyramid has a base with area B square units and a height of h units. What is the volume V of the pyramid?

ACTIVITY 11.4.2   Surface Area and Volume

Material: sharp pencil and metric ruler   Name _____

The purpose of this activity is to apply the formulas for the surface area of a regular prism (each base is regular polygon and the faces are rectangles), a regular pyramid (the base is a regular polygon and the faces are isosceles or equilateral triangles), and cylinder. Students will use a metric ruler to obtain the measurements needed in the formulas for the surface area and volume of prisms, pyramids, and cylinders.

1.   The diagram below shows the net of a pentagonal prism. Use a metric ruler to find each measurement.
     Round the final answer to the nearest mm, square mm, or cubic mm.
     a. length of a side of the base
     b. apothem of a base
     c. area of a base
     d. height of the prism
     e. lateral surface area of the prism
     f. surface area of the prism
     g. volume of the prism

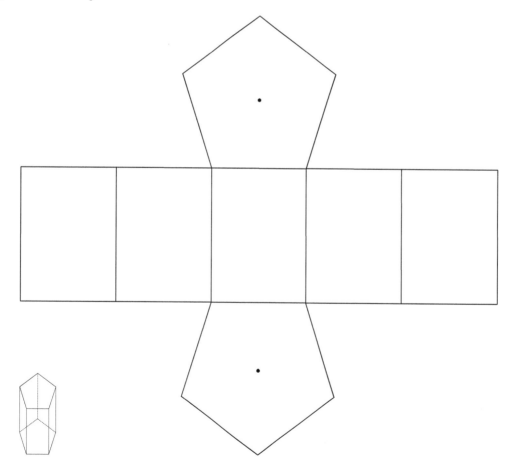

2. The diagram below shows the net of an octagonal pyramid. Use a metric ruler to find each measurement.

   Round the final answer to the nearest whole unit (mm, square mm, or cubic mm).

   a. length of a side of the base

   b. apothem of the base

   c. area of the base

   d. slant height of the pyramid

   e. lateral surface area of the pyramid

   f. surface area of the pyramid

   g. height of the pyramid (Hint: apply the Pythagorean Theorem)

   h. volume of the pyramid

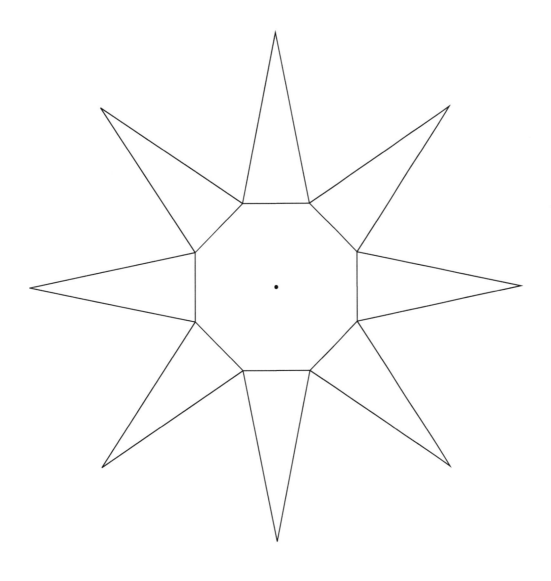

3.   The diagram below shows the net of a cylinder. Use a metric ruler to find each measurement.

Round the final answer to the nearest whole unit (mm, square mm, or cubic mm).

   a. radius of a base

   b. area of a base

   c. height of the cylinder

   d. lateral surface area of the cylinder

   e. surface area of the cylinder

   f. volume of the cylinder

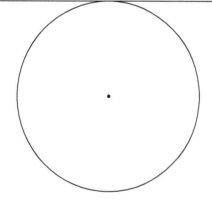

ACTIVITY 11.4.3   Dimensional Analysis and Volume

Material: sharp pencil   Name _____

The purpose of this activity is to gain practice converting volume measurements using dimensional analysis.

1. The following cube represents 1 cubic feet.

    a. What are the dimensions of the cube?

    b. Explain why the following cube represents 1 cubic yard.

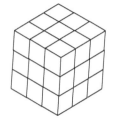

    c. How many cubic feet are in 1 cubic yard?

2. Now we will demonstrate how to use dimensional analysis to convert 1 cubic yard to cubic feet:

$$1\text{ yd}^3 = 1\text{ yd}^3 \times \left(\frac{3\text{ ft}}{1\text{ yd}}\right)^3 = 1\text{ yd}^3 \times \frac{3^3\text{ ft}^3}{1^3\text{ yd}^3} = 1\ \cancel{\text{yd}^3} \times \frac{27\text{ ft}^3}{1\ \cancel{\text{yd}^3}} = 27\text{ ft}^3.$$

Use dimensional analysis to convert the measurements. Round the final answer to the nearest 0.1. (Note: 1 inch = 2.54 cm)

    a. $4\text{ ft}^3 =$ _____ in^3

    b. $8\text{ ft}^3 =$ _____ cm^3

    c. $150\text{ ft}^3 =$ _____ m^3

3. Use dimensional analysis to convert the measurements. Round the final answer to the nearest 0.1.

Measurement unit	Relationship
clip	7 clips = 4 pens
pen	
zap	5 zaps = 3 clips

    a. 12 cubic clips = _____ cubic zaps

    b. 18 cubic pens = _____ cubic clips

    c. 14 cubic pens = _____ cubic zaps

    d. 22 cubic zaps = _____ cubic clips

ACTIVITY 12.1.1 Triangle Congruence Axioms

Material: sharp pencil, protractor, scissors,   Name _____

tape, and line segment arrays (p. A16)

The definition of congruent triangles lists six conditions that two triangles have to satisfy to have the same shape and size. An *axiom* is a statement that we accept as true. Triangle axioms for two triangles allow us, in some cases, to check three of the six conditions rather than all six conditions. Triangle axioms are important because we use them to establish properties of quadrilaterals, such as parallelograms, rectangles, squares, rhombuses, and kites. This activity makes it easier to see that the triangle axioms are reasonable statements to accept.

Definition: $\triangle ABC$ is congruent to $\triangle DEF$, denoted $\triangle ABC \cong \triangle DEF$, if and only if
$\overline{AB} \cong \overline{DE}$, $\overline{BC} \cong \overline{EF}$, $\overline{AC} \cong \overline{DF}$, $\angle A \cong \angle D$, $\angle B \cong \angle E$, and $\angle C \cong \angle F$.

Using words, two triangles are **congruent** (same size and same shape) if and only there is a correspondence between the vertices such that all pairs of corresponding sides are congruent and all pairs of corresponding angles are congruent. According to the definition, we would need to check six conditions. Are all six conditions necessary? It turns out that in some cases, you just have to check three conditions.

1. A teacher writes the congruence statement $\triangle OLD \cong \triangle NEW$.

    a. Draw a picture that represents the congruence.

    b. List all corresponding line segments and angles.

Before continuing, refer to Figures 1-3 in Activity 11.3.2 for connecting the line segment arrays to form triangles.

2. A triangle has sides of length 6 units, 7 units, and 10 units. Use scissors to cut three line segments of the given lengths from the line segment arrays on page A16.

    a. Form a triangle with the line segments. Use tape to preserve the shape of the triangle.

    b. Compare your triangle to a triangle formed by another student. Are the two triangles congruent?

    c. Based on your results in part (b), do you need to check all six conditions of the definition to check if two triangles are congruent?

    d. Does the diagram below provide enough information to conclude the triangles are congruent? If so, write a congruence statement.

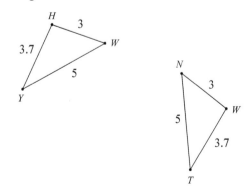

3. Carlos drew an isosceles triangle. Then he drew a line segment from the apex U of the isosceles triangle to the midpoint M of the base $\overline{RN}$ of the triangle.

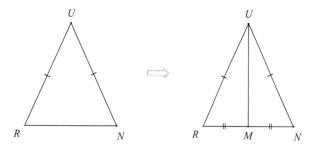

a. Carlos claimed that $\triangle\,RUM$ is congruent to $\triangle\,NUM$. Explain why Carlos is correct.

b. Carlos then concluded that the base angles $\angle R$ and $\angle N$ of the isosceles triangle are congruent. Explain why Carlos is correct.

4. The diagram shows a circle with center S and points P, Q, and T on the circle. Use logical reasoning to determine the measures of the numbered angles.

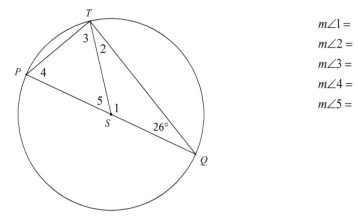

$m\angle 1 =$

$m\angle 2 =$

$m\angle 3 =$

$m\angle 4 =$

$m\angle 5 =$

5. A triangle has two sides of length 5 units and 7 units. The angle formed by these two sides (called the *included angle*), is 40°. Use scissors to cut two line segments of the given lengths from the line segment arrays on page A16.

a. Use a protractor to draw an angle with measure 40°. Place the line segments of length 5 units and 7 units along the sides of the angle.

b. Use the scissors to cut another line segment array to form a triangle. Use tape to preserve the shape of the triangle.

c. Compare your triangle to a triangle formed by another student. Are the two triangles congruent?

d. Does the diagram below provide enough information to conclude the triangles are congruent? If so, write a congruence statement.

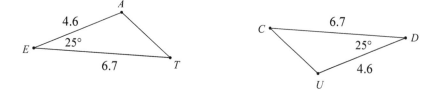

6. A triangle has two angles with measures 30° and 70°. The length of the side that joins the vertices of these two angles (called the *included side*) has length 7 units. Use scissors to cut the line segment of the given length from the line segment arrays on page A16.

    a. Use a protractor to draw two angles with measures 30° and 70° such that the distance between the vertices of the angles is 7 units. Place the line segment of length 7 units along the included side.

    b. Use scissors to cut two more line segment arrays that can be used to form a triangle with the two angles and the included side. Use tape to preserve the shape of the triangle.

    c. Compare your triangle to one formed by another student. Are the two triangles congruent?

    d. Does the diagram below provide enough information to conclude the triangles are congruent? If so, write a congruence statement.

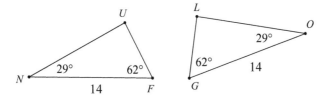

7. $\triangle PQR$ has the following properties: $m\angle P = 30°$, $m\angle Q = 110°$, and $QR = 7$ units. Note that $\overline{QR}$ is referred to as a non-included side for $\angle P$ and $\angle Q$.

    a. Use scissors to cut a line segment of length 7 units from the line segment arrays on page p. A16.

    b. Use a straightedge and protractor to draw $\angle Q$ such that $m\angle Q = 110°$. Then align the line segment of length 7 units on one side to locate point R.

    c. If $m\angle P = 30°$ and $m\angle Q = 110°$, then what is $m\angle R$?

    d. Draw a line from R to the other side of $\angle Q$, and let P be the intersection of the two lines. Make sure that such that $\angle P$ has measure 30°.

    e. Compare your triangle to one formed by another student. Are the two triangles congruent?

    f. Does the diagram below provide enough information to conclude the triangles are congruent? If so, write a congruence statement.

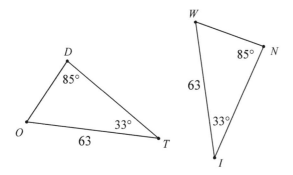

ACTIVITY 12.2.1   Constructing a Triangle

Material: sharp pencil, compass, and straightedge   Name _____

Suppose you are given three line segments of lengths a, b, and c units that form a triangle. How can you use a compass and straightedge to construct the triangle? This activity provides directions for constructing the triangle.

1.   Follow the directions in part (a)-(e) to construct a triangles with sides of lengths a, b, and c units.

a

b

c

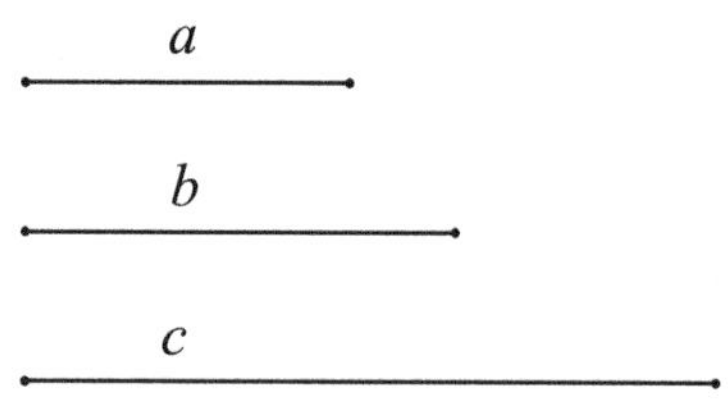

a.   Use the straightedge to construct a line.

b.   Pick a point on the line and label it P. Open the compass so that it has radius a units. Put the compass point at P create an arc so that it intersects the line. Label the point of intersection Q.

c.   Open the compass so that it has radius b units. Construct a circle with center P.

d.   Open the compass so that it has radius c units. Construct a circle with center Q.

e.   The two circles intersect at two points. Label one of the points of intersection R. Use a straightedge to construct the sides of $\triangle PQR$.

2.   Apply the steps described in Problem 1 to construct an isosceles triangle with legs having length a units and the base having length b units.

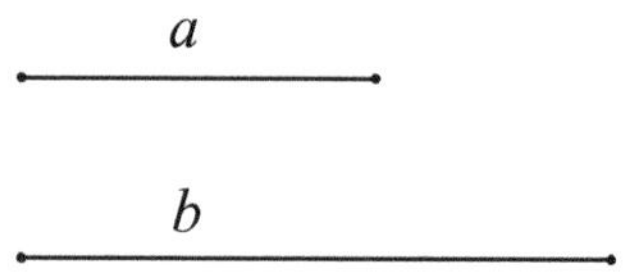

a

b

3.   Apply the steps described in Problem 1 to construct an equilateral triangle with sides congruent to the line segment shown.

b

4.   The following line segments cannot be used to form a triangle. Apply the steps described in Problem 1 to see what happens.

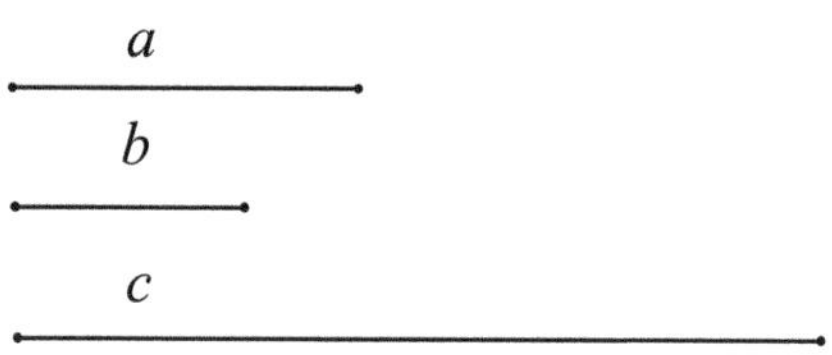

a

b

c

ACTIVITY 12.2.2 Constructing a Regular Hexagon

Material: sharp pencil, compass, and straightedge   Name _____

A regular *n*-gon is a polygon with *n* sides such that the sides have the same length and the interior angles have the same angle measure. This activity leads you through the steps to construct a regular hexagon ($n = 6$).

1. We will describe how to use a compass and straightedge to construct a regular hexagon.

    a. **Step 1.** Use the compass to construct a circle (with radius *r* units, where *r* is the distance between the compass point and the center point of the compass). Then use a straightedge to locate a point on the circle. Label the point of intersection *A*.

    b. **Step 2.** Maintaining the same compass opening (so that the distance between the center point and pencil point of the compass is the radius of the circle), place the center point of the compass at *A* and swing an arc so that it intersects the circle. Label the point of intersection *B*.

    c. **Step 3.** Maintaining the same compass opening, place the center point of the compass at *B* and swing an arc so that it intersects the circle. Label the point of intersection *C*. Repeat this step to label the points *D*, *E*, *F*, and *G*.

    d. **Step 4.** Use the straightedge to construct the hexagon *ABCDEF*.

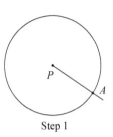
Step 1

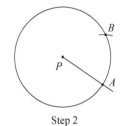
Step 2

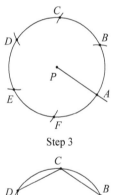
Step 3

2. Now you have constructed hexagon *ABCDEF*. Let *r* be the radius of the circle.

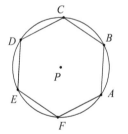

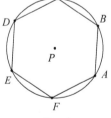

Step 4

    a. What is the distance between *P* and *A*? Explain.

    b. What is the distance between *P* and *B*? Explain.

    c. What is the distance between *A* and *B*?

    d. Use a straightedge to construct △ *APB* and △ *BPC*.

    e. Explain why △ *APB* ≅ △ *BPC*.

    f. What is the measure of each interior angle of △ *APB*?

    g. What type of triangle is △ *APB*?

    h. Explain why the sides of hexagon *ABCDEF* have the same length.

    i. Explain why the measure of each interior angle of hexagon *ABCDEF* is the same.

    j. Explain why *ABCDEF* is a regular hexagon.

3.   How could you use the construction of a hexagon as the basis for constructing a regular dodecagon ($n = 12$) with vertices on the circle?

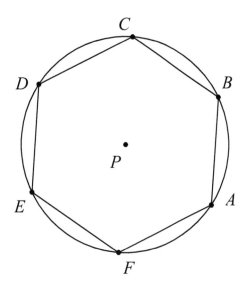

ACTIVITY 12.3.1   Similar Triangles

Material: sharp pencil and straightedge   Name _____

Similarity is a relationship between two geometric figures that have the same shape and possibility different sizes. For polygons, similarity specifies specific relationships between the corresponding sides, as well as specific relationships between the corresponding angles.

1.   *Similar triangles* have the same shape. Basically, the larger triangle looks like a magnification of the other.

   $\triangle ABC$ and $\triangle DEF$ in the diagram below (on cm dot paper) are similar triangles. The vertices have the following correspondences: $A \leftrightarrow D$, $B \leftrightarrow E$, and $C \leftrightarrow F$.

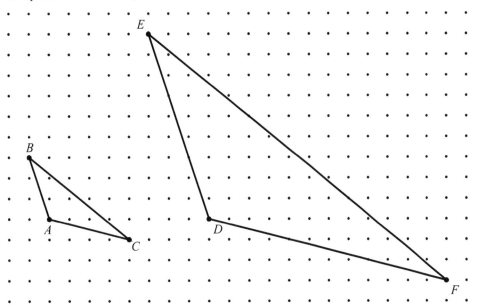

   a.   Use a protractor to measure the angles.

   $m\angle A =$ _____ , $m\angle B =$ _____ , $m\angle C =$ _____

   $m\angle D =$ _____ , $m\angle E =$ _____ , $m\angle F =$ _____

   b.   Fill in the blanks with $=$, $<$, or $>$ to compare the corresponding angles.

   $m\angle A$ ____ $m\angle D$, $m\angle B$ ____ $m\angle E$, $m\angle C$ ____ $m\angle F$

   c.   Use the Pythagorean Theorem to determine the length of each side of $\triangle ABC$:

   $AB =$ ____$\sqrt{10}$____ cm , $BC =$ _____ cm , $AC =$ _____ cm .

   d.   Use the Pythagorean Theorem to determine the length of each side of $\triangle DEF$:

   $DE =$ _____ cm , $EF =$ ____$\sqrt{369}$____ cm , $DF =$ _____ cm .

   e.   Determine the ratios of the corresponding sides (quotients): $\dfrac{DE}{AB} =$ ____ , $\dfrac{EF}{BC} =$ ____ , $\dfrac{AC}{DF} =$ ____ .

   f.   How are the corresponding angles related?

   g.   How are the corresponding sides related?

2. Refer to the definition of similar triangles in your textbook in Section 12.3, and modify the definition to write the definition of the similarity statement $\triangle FGH \sim \triangle PQR$.

   a. Use words.

   b. Use a diagram.

   c. Use mathematical notation.

3. $\triangle CAT \sim \triangle FUR$. Find the missing lengths.

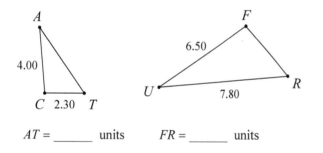

   $AT =$ _____ units $FR =$ _____ units

4. You are given $\triangle STU \sim \triangle PQR$ and $\dfrac{ST}{PQ} = 20$.

   a. Suppose $ST = 140$ cm. Find PQ.

   b. Suppose $SU = 14$ cm. Find PR.

   c. Suppose $QR = 32$ cm. Find TU.

5. Use a straightedge to draw a triangle.

    a. Identify the largest angle of the triangle.

    b. Identify the longest side of the triangle.

    c. How are the positions of the largest angle and longest side related?

    d. Identify the smallest angle of the triangle.

    e. Identify the shortest side of the triangle.

    f. How are the positions of the smallest angle and shortest side related?

6. Suppose $\triangle LMS$ and $\triangle PQR$ are scalene triangles, $\triangle LMS \sim \triangle PQR$, and $\triangle LMS$ has largest angle $\angle L$ and smallest angle $\angle S$.

    a. Represent $\triangle LMS \sim \triangle PQR$ with a diagram.

    b. What is the largest angle of $\triangle PQR$?

    c. Explain why the longest sides of the triangles must be corresponding sides.

    d. What is the smallest angle of $\triangle PQR$?

    e. Explain why the shortest sides of the triangles must be corresponding sides.

7. The two scalene triangles shown are similar triangles. Write a similarity statement.

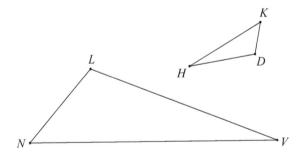

8. Two triangles are similar. One triangle has vertices K, L, and M with sides having lengths $KL = 63$ cm, $LM = 27$ cm, and $KM = 36$ cm. The other triangle has vertices S, T, and U with sides having lengths $TU = 48$ cm, $US = 84$ cm, and $ST = 36$ cm. Write a similarity statement for the similar triangles.

ACTIVITY 12.3.2   Relating the Perimeters and Areas of Similar Triangles

Material: sharp pencil   Name _____

We already know that if two triangles are similar, then there is a correspondence between the vertices such that corresponding angles are congruent and the corresponding sides have identical ratios (quotients). In this activity, we will explore the relationship between (a) the perimeters of two similar triangles, and (b) the areas of two similar triangles.

1. $\triangle DEF$ and $\triangle ABC$ are similar triangles. The unit of measurement for length is cm.

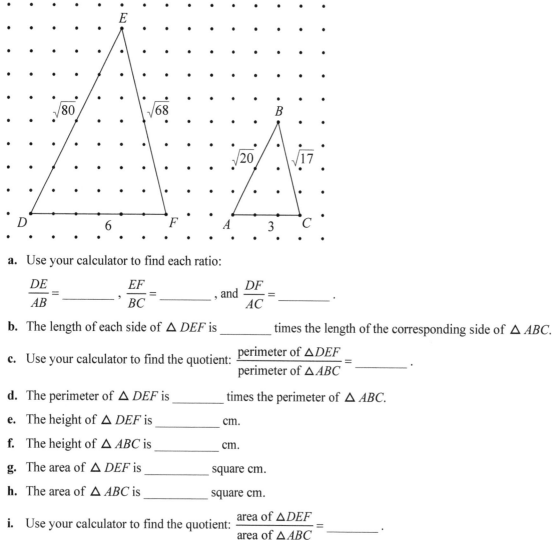

   a. Use your calculator to find each ratio:

   $\dfrac{DE}{AB} =$ _____ , $\dfrac{EF}{BC} =$ _____ , and $\dfrac{DF}{AC} =$ _____ .

   b. The length of each side of $\triangle DEF$ is _____ times the length of the corresponding side of $\triangle ABC$.

   c. Use your calculator to find the quotient: $\dfrac{\text{perimeter of } \triangle DEF}{\text{perimeter of } \triangle ABC} =$ _____ .

   d. The perimeter of $\triangle DEF$ is _____ times the perimeter of $\triangle ABC$.

   e. The height of $\triangle DEF$ is _____ cm.

   f. The height of $\triangle ABC$ is _____ cm.

   g. The area of $\triangle DEF$ is _____ square cm.

   h. The area of $\triangle ABC$ is _____ square cm.

   i. Use your calculator to find the quotient: $\dfrac{\text{area of } \triangle DEF}{\text{area of } \triangle ABC} =$ _____ .

   j. The area of $\triangle DEF$ is _____ times the area of $\triangle ABC$.

2. $\triangle DEF$ and $\triangle ABC$ are similar triangles. The unit of measurement for length is cm.

a. Verify that $DE = \sqrt{45}$ cm and $AB = \sqrt{5}$ cm.

b. Verify that $EF = \sqrt{180}$ cm and $BC = \sqrt{20}$ cm.

c. Find DF and AC: $DF =$ _____ cm and $AC =$ _____ cm.

d. Use your calculator to find each ratio:

$\dfrac{DE}{AB} =$ _____ , $\dfrac{EF}{BC} =$ _____ , and $\dfrac{DF}{AC} =$ _____ .

e. The length of each side of $\triangle DEF$ is _____ times the length of the corresponding side of $\triangle ABC$.

f. Use your calculator to find the quotient: $\dfrac{\text{perimeter of } \triangle DEF}{\text{perimeter of } \triangle ABC} =$ _____ .

g. The perimeter of $\triangle DEF$ is _____ times the perimeter of $\triangle ABC$.

h. The height of $\triangle DEF$ is _____ cm.

i. The height of $\triangle ABC$ is _____ cm.

j. The area of $\triangle DEF$ is _____ square cm.

k. The area of $\triangle ABC$ is _____ square cm.

l. Use your calculator to find the quotient: $\dfrac{\text{area of } \triangle DEF}{\text{area of } \triangle ABC} =$ _____ .

m. The area of $\triangle DEF$ is _____ times the area of $\triangle ABC$.

3. Suppose $\triangle STU \sim \triangle PQR$ and $\dfrac{ST}{PQ} = 12$. Fill in the blanks.

a. The length of each side of $\triangle STU$ is _____ times the length of the corresponding side of $\triangle PQR$.

b. The perimeter of $\triangle STU$ is _____ times the perimeter of $\triangle PQR$.

c. The area of $\triangle STU$ is _____ times the area of $\triangle PQR$.

4. Suppose $\triangle STU \sim \triangle PQR$ and $\dfrac{ST}{PQ} = 20$. Fill in the blanks.

a. The length of each side of $\triangle STU$ is _____ times the length of the corresponding side of $\triangle PQR$.

b. The perimeter of $\triangle STU$ is _____ times the perimeter of $\triangle PQR$.

c. The area of $\triangle STU$ is _____ times the area of $\triangle PQR$.

ACTIVITY 12.3.3   Motivating and Applying Similarity Axioms

Material: sharp pencil, protractor, and straightedge   Name _____

The purpose of this activity is to

(a) motivate the AA Similarity Axiom for triangles,

(b) apply the AA, SSS, and SAS Similarity Axioms to determine an unknown measurement using known measurements of similar triangles, and

(c) learn how to sketch a triangle that is similar to a given triangle.

In Section 12.1, we learned the definition of congruent triangles is based on six angle measurements and six side measurements. Then we learned that in some cases, we only needed some of the measurements to conclude that two triangles are congruent (think: SSS Congruence Axiom, SAS Congruence Axiom, ASA Congruence Axiom, and AAS Congruence Axiom). In Section 12.3, we learned that the definition of similar triangles is based on six angle measurements and six side measurements. To verify $\triangle ABC \sim \triangle DEF$, we would need to find the measurements $m\angle A$, $m\angle B$, $m\angle C$, $m\angle D$, $m\angle E$, $m\angle F$, AB, BC, AC, DE, EF, and DF, then verify $m\angle A = m\angle D$, $m\angle B = m\angle E$, $m\angle C = m\angle F$, and then verify the quotients $\frac{AB}{DE}$, $\frac{BC}{EF}$, $\frac{AC}{DF}$ are identical. Are all the measurements necessary to prove that two triangles are similar?

1. Do the following.

    a. The diagram shows $\triangle ABC$ with two angles having measurements $75°$ and $40°$. Use a protractor and straightedge to draw $\triangle DEF$ having one angle with measurement $75°$ and another angle having measurement $40°$. Make sure $\triangle ABC$ and $\triangle DEF$ different sizes.

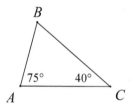

    b. Measure the sides of each triangle and complete the table. Round the quotients to the nearest 0.1.

$\triangle ABC$	$\triangle DEF$	corresponding parts	relationship
$m\angle A = 75°$	$m\angle D =$ _____	$\angle A$ and $\angle D$	
$m\angle B =$ _____	$m\angle E =$ _____	$\angle B$ and $\angle E$	
$m\angle C = 40°$	$m\angle F =$ _____	$\angle C$ and $\angle F$	
$AB =$ _____ mm	$DE =$ _____ mm	$\overline{AB}$ and $\overline{DE}$	$\dfrac{DE}{AB} =$ _____
$BC =$ _____ mm	$EF =$ _____ mm	$\overline{BC}$ and $\overline{EF}$	$\dfrac{EF}{BC} =$ _____
$AC =$ _____ mm	$DF =$ _____ mm	$\overline{AC}$ and $\overline{DF}$	$\dfrac{DF}{AC} =$ _____

    c. Are the corresponding angles equal?

    d. Are the ratios (quotients) of corresponding sides equal?

e. Ideally, if there were no drawing errors and no measurement errors, then the corresponding angles would be congruent and the three quotients would be identical. This motivates the **Angle-Angle Similarity Axiom** (*AA Similarity Axiom*): "If two angles in a triangle are congruent to the corresponding angles in another triangle, then the two triangles are similar." Refer to Section 12.3 in your textbook. Draw a diagram that represents the AA Similarity Axiom.

f. Refer to Section 12.3 in your textbook. Write the SAS Similarity Axiom. Then draw a diagram that represents the SAS Similarity Axiom.

g. Refer to Section 12.3 in your textbook. Write the SSS Similarity Axiom. Then draw a diagram that represents the AA Similarity Axiom.

2.   The diagram shows two triangles.

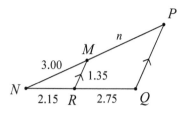

    a.   Explain why the two triangles are similar.

    b.   Write a similarity statement.

    c.   Find n (round to the nearest 0.01 unit).

4.   The diagram shows two similar triangles.

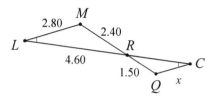

    a.   Write a similarity statement.

    b.   Explain why they are similar.

    c.   Find x (round to the nearest 0.01 unit).

3.   One triangle has vertices K, L, and M with sides having lengths $KL = 63$ cm, $LM = 27$ cm, and $KM = 36$ cm. The other triangle has vertices S, T, and U with sides having lengths $TU = 48$ cm, $US = 84$ cm, and $ST = 36$ cm.

    a.   Explain why the two triangles are similar.

    b.   Write a similarity statement.

5.   Explain why the two triangles are similar.

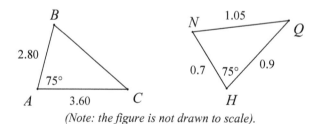

(Note: the figure is not drawn to scale).

ACTIVITY 12.3.4 Sketching Similar Triangles

Material: sharp pencil and straightedge   Name _____

Please review Concept Map 12.1 in Section 12.3, which illustrates three ways to draw a triangle that is similar to a given triangle. Draw markings such as angle arcs or arrowheads as appropriate.

1. Locate a point E so that $\triangle ABC \sim \triangle DEC$.

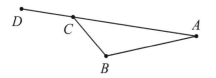

2. Locate a point E so that $\triangle ABC \sim \triangle DEC$.

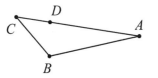

3. Locate a point E so that $\triangle ABC \sim \triangle AED$.

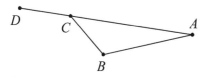

4. Draw a triangle that is similar to the given triangle. Use three different strategies.

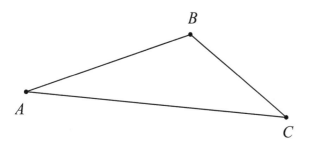

Activity 13.1.1 Representing Lines and Defining Slope

Materials: Sharp pencil and straightedge

Name _____

There are three types of lines: vertical lines, horizontal lines, and oblique lines. We can represent the lines using tables, graphs, and algebra. Here's how we represent the three types of lines using algebra:

type	algebra	conditions
vertical line	$x = a$	a is any number
horizontal line	$y = b$	b is any number
oblique line	$y = mx + b$	$m \neq 0$ and b is any number

1. Fill in the blanks: $(0, 0)$, x-axis, x-coordinate, y-axis, y-coordinate, coordinate plane, ordered pair, origin.

   a. A _____ (also called the **Cartesian plane**) is a plane that contains two perpendicular number lines.

   b. Their intersection of the two number lines is called the _____.

   c. The coordinate plane lets us prescribe the *location* of a point using two numbers called an _____ that has the form (a, b).

   d. The first number a, commonly called the _____, tells you how far to move left or right from the vertical axis, and the second number b, commonly called the _____, tells how far to move above or below from the horizontal axis.

   e. The ordered pair _____ is the location of the origin.

A basic postulate for points and lines is: *Given two points, there is exactly one line containing the two points*. To graph a line, we just need to find two points that belong to the line and then use a straightedge to draw the line through the points.

2. Do the following.

   a. Make a table for the equation $x = 4$.

   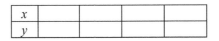

   b. Draw the graph of the equation $x = 4$.

   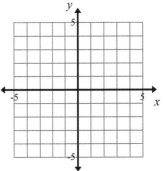

3. Do the following.

   a. Make a table for the equation $y = 3$.

   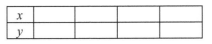

   b. Draw the graph of the equation $y = 3$.

   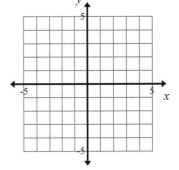

4. Do the following.
    a. Make a table for the equation $y = 3x - 1$.

    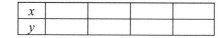

    b. Draw the graph of the equation $y = 3x - 1$.

    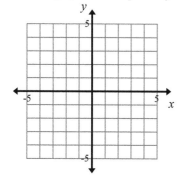

The **slope** (steepness) of a line is the number of units the graph rises or falls vertically for every one unit you move to the right (in the same direction that we read a number line, from left to right). This means that the slope is equal to " $\frac{\text{rise}}{\text{run}}$." For example, the slope of line u is $\frac{\text{rise}}{\text{run}} = \frac{3}{2}$ and the slope of line v is $\frac{\text{rise}}{\text{run}} = \frac{-2}{5}$.

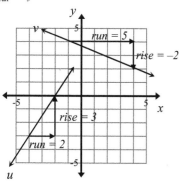

5. Find the slopes of lines m and n.

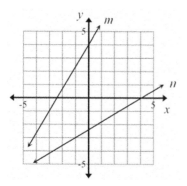

6. Find the slopes of lines p and q.

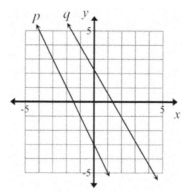

7. Graph the line with slope $\frac{-4}{3}$ and passing through the point $A(4,-1)$.

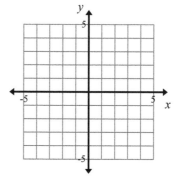

8. Graph the line with slope $\frac{3}{2}$ and passing through the point $A(-2,-4)$.

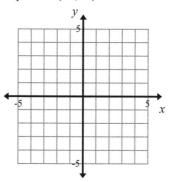

9. Graph the line with slope $\frac{3}{4}$ and passing through the point $A(4,3)$.

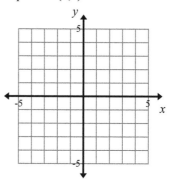

Activity 13.1.2 Slope-Intercept Form of an Oblique Line

Material: sharp pencil   Name _____

This activity focuses on (a) using algebra to determine the slope of an oblique line, (b) determining the slope and *y*-intercept of an oblique line written in *slope-intercept form*, and (c) writing the equation of a line passing through two given points.

- The **slope** (steepness) of a nonvertical line through points (x_1, y_1) and (x_2, y_2) is the quotient

$$slope = \frac{y_2 - y_1}{x_2 - x_1}.$$

- The ***x*-intercept** of a nonhorizontal line is the ordered pair $(k, 0)$ where the line intersects the *x*-axis.

- The ***y*-intercept** of a nonvertical line is the ordered pair $(0, k)$ where the line intersects the *y*-axis.

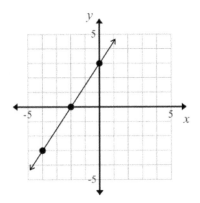

1. Refer to the graph above.
   a. Pick any two points on the line and apply the formula for the slope.
   b. Repeat part (a) using a different pair of points.
   c. Does the slope of the line depend on the two points selected?
   d. What is the *x*-intercept of the line?
   e. What is the *y*-intercept of the line?

2. The equation of a line is $y = 4x - 7$.
   a. Calculate the slope of the line by finding two points (x_1, y_1) and (x_2, y_2) on the line and then using the slope formula.
   b. Calculate the *x*-intercept of the line.
   c. Calculate the *y*-intercept of the line.

3. The equation of a line is $y = -5x + 2$.
   a. Calculate the slope of the line by finding two points (x_1, y_1) and (x_2, y_2) on the line and then using the slope formula.
   b. Calculate the *x*-intercept of the line.
   c. Calculate the *y*-intercept of the line.

4. Look for patterns in Problems 2 and 3 to determine the slope and *y*-intercept of each oblique line.
   a. $y = 4x + 5$
   b. $y = -3x - 2$

5. The *slope-intercept form* of an oblique line is $y = mx + b$. What information about the line does this form reveal?

6. The equation of an oblique line with slope 4 can be written in the form $y = 4x + b$. Suppose the line passes through the point $A(3, 5)$. Explain how to find b.

7. A line passes through $A(5, -2)$ and $B(1, 3)$.
   a. Find the slope of the line.
   b. Write the equation of the line in the slope-intercept form.

8. A line passes through $A(2, 5)$ and $B(6, 8)$.
   a. Find the slope of the line.
   b. Write the equation of the line in the slope-intercept form.

9. Determine the x- and y-intercepts of the line shown.

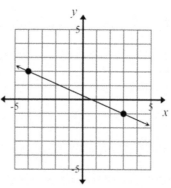

10. A study focused on the speed v (in miles per hour) of a motorcycle and the total time t (in seconds) it takes for the motorcycle to reach a complete stop. At 25 mph, the stopping time for the motorcycle is 1.95 sec. At 60 mph, the stopping time for the motorcycle is 3.70 sec. The speed and stopping time are linearly related.
   a. Express the stopping time t in terms of the speed v.
   b. Interpret the ordered pair (45, 2.95).
   c. What is the stopping time when the vehicle is traveling at a speed of 70 mph?

11. Find two points (x_1, y_1) and (x_2, y_2) on the vertical line $x = 3$. What happens when you apply the slope formula to these points? What is the slope of a vertical line?

Activity 13.1.3 Relating Slopes of Perpendicular and Parallel Oblique Lines

Material: sharp pencil and straightedge   Name _____

This activity focuses on relationships between perpendicular and parallel lines. How are the slopes of perpendicular oblique lines related? How are the slopes of parallel oblique lines related?

1. The figure shows two oblique lines p and q in the same coordinate plane.

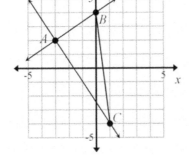

a. Calculate the lengths of the sides of $\triangle ABC$.

b. The converse of the Pythagorean Theorem states that if a triangle with sides of length a, b, and c units satisfies the equation $a^2 + b^2 = c^2$, then the triangle is a right triangle. Verify that $\triangle ABC$ is a right triangle.

c. Explain why lines p and q are perpendicular.

d. Determine the slopes of lines p and q.

e. Verify that the product of the slopes of lines p and q is -1.

f. Complete the statement: *Two oblique lines are perpendicular* if and only if ...

2. The figure shows two oblique lines p and q in the same coordinate plane.

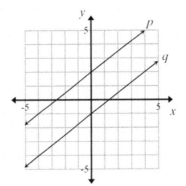

a. In Section 12.1, we learned that if the opposite sides of a quadrilateral are congruent, then the quadrilateral is a parallelogram. Form quadrilateral $ABCD$ such that A and B belong to line p, C and D belong to line q. Pick the vertices of quadrilateral $ABCD$ strategically such that

(i) you can determine the lengths of the sides of $ABCD$ and

(ii) the opposite sides of your quadrilateral are congruent.

b. Explain why lines p and q are parallel.

c. Determine the slopes of lines p and q.

d. Compare the slopes of lines p and q.

e. Complete the statement: *Two oblique lines are parallel* if and only if ...

3. Two lines are parallel. The slope of one line is $\frac{4}{5}$. What is the slope of the other line?

7. Two lines are perpendicular. The slope of one line is $\frac{3}{7}$. What is the slope of the other line?

4. Draw a line that is parallel to the given line l and passes through the given point. (Hint: Locate another point on line k, and then use a straightedge to draw the line).

8. Draw a line that is perpendicular to the given line l and passes through the given point. (Hint: Locate another point on line k, and then use a straightedge to draw the line).

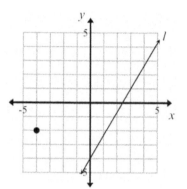

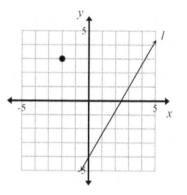

5. The equation of line l is $y = 5x + 3$. Write the equation of the line parallel to l and passing through $(2, -3)$.

9. The equation of line l is $y = 5x + 3$. Write the equation of the line perpendicular to l and passing through $(2, -3)$.

6. The equation of line l is $y = \frac{4}{5}x + 2$. Write the equation of the line parallel to l and passing through $(20, 14)$.

10. The equation of line l is $y = \frac{4}{5}x + 2$. Write the equation of the line perpendicular to l and passing through $(20, 8)$.

Activity 13.2.1 Quadrilaterals and Coordinate Geometry

Material: sharp pencil   Name _____

Coordinate geometry uses algebra to study geometric properties and relationships. It is useful for verifying properties of polygons such as parallelogram and rectangles. In coordinate geometry, each vertex of a polygon corresponds to an ordered pair. As you know, a parallelogram is a quadrilateral such that the opposite sides of the quadrilateral are parallel.

In this activity you will learn how to use coordinate geometry to verify a property of parallelograms (the opposite sides of a parallelogram are congruent) and rectangles (the diagonals are congruent). A rectangle is a special type of parallelogram, so the opposite sides of a rectangle are congruent, too. The distance and slope formulas play critical roles in this activity. The **distance** AB between points $A(x_1, y_1)$ and $B(x_2, y_2)$ in the coordinate plane is

$$AB = \sqrt{(x_2 - x_1)^2 + (y_2 - y_1)^2} \text{ , and the } \textbf{slope} \text{ of the line containing } A(x_1, y_1) \text{ and } B(x_2, y_2) \text{ is } slope = \frac{y_2 - y_1}{x_2 - x_1}.$$

1. Refresher: Points A and B are shown.

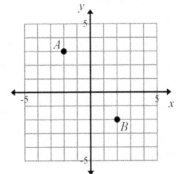

   a. Find the distance between A and B.

   b. Find the slope of the line containing A and B.

2. In this problem, you will learn how to place a parallelogram in the coordinate plane and (i) find the coordinates of its vertices, and (ii) prove that the opposite sides of the parallelogram are congruent. To simplify the analysis, we will place the parallelogram in the coordinate plane strategically to introduce as many zeros in the coordinates of the vertices as possible and then use variables as needed for the vertices of the parallelogram. Suppose parallelogram $ABCD$ has height b units and base c units. We can place vertex A at the origin $(0, 0)$ and vertex D at $(c, 0)$. $\overline{AD}$ would be the base with length c units. The y-coordinate of B must be b units. There is no restriction on the x-coordinate of B, so we assign it the variable a. Then B has coordinates (a, b).

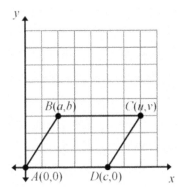

The following questions will lead you along so that you can determine the coordinates of C so that the opposite sides of the quadrilateral are parallel. Then $ABCD$ will be a parallelogram. Then you will prove that the opposite sides of the parallelogram are congruent. Please follow along.

    a. $\overrightarrow{AD}$ is a horizontal line. What is the slope of $\overrightarrow{AD}$?

    b. $\overrightarrow{BC}$ and $\overrightarrow{AD}$ are parallel. What is the slope of $\overrightarrow{BC}$?

    c. Apply the formula $slope = \dfrac{y_2 - y_1}{x_2 - x_1}$ to the points $B(a,b)$ and $C(u,v)$ to determine the y-coordinate v of C.

    Be sure to use the value of the slope that you found in part (b).

    d. $\overrightarrow{AB}$ is an oblique line. Find an expression for the slope of $\overrightarrow{AB}$.

    e. $\overrightarrow{AB}$ and $\overrightarrow{CD}$ are parallel. What is the slope of $\overrightarrow{CD}$?

    f. Apply the formula $slope = \dfrac{y_2 - y_1}{x_2 - x_1}$ to the points $C(u,v)$ and $D(c,0)$ to determine the x-coordinate u of C.

    Be sure to use the value of the slope that you found in part (e) and the value of v that you found in part (c).

    g. Now you know the coordinates of the vertices of the parallelogram. Use the coordinates to calculate AB, BC, CD, and AD.

    h. Compare AB and CD.

    i. Compare BC and AD.

    j. Write a statement that tells how the lengths of the opposite sides of a parallel are related.

3. In this problem, we would like to prove that the diagonals of a rectangle are congruent. Are you ready?

A rectangle has length l units and width w units. Locate the rectangle $ABCD$ strategically in the coordinate plane so that you introduce as many zeros as possible in the coordinates of the vertices of the rectangle. Be sure to take into account the given information (length and width) when you use variables for the vertices, take into account the definition that a rectangle is special type of parallelogram, and take into account that we've already proven that the opposite sides of a parallelogram are congruent. After the vertices have been labeled with letters and their coordinates, you are ready to prove that the diagonals of a rectangle are congruent by showing that AC equals BD.

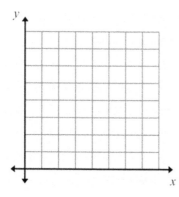

Activity 13.2.2 More Experience with Coordinate Geometry

Material: sharp pencil   Name _____

1. $\triangle HAT$ is an isosceles triangle with base $\overline{HT}$ and legs $\overline{AH}$ and $\overline{AT}$. Suppose the length of the base is b units and the height of the triangle is h units. The base of the triangle is placed along the horizontal axis as shown. Then H has coordinates $(0, 0)$ and T has coordinates $(b, 0)$. The y-coordinate of the apex A is h since the triangle has height h units. There is no given restriction on the x-coordinate of A, so let a be the x-coordinate of A. The result is shown.

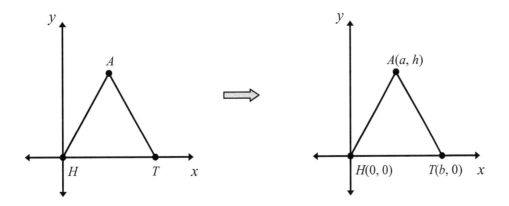

a. Use the fact that $AH = AT$ to verify that the x-coordinate of A is can be expressed $a = \frac{b}{2}$.

b. Find the coordinates of the midpoint M of the base $\overline{HT}$.

c. What is the slope of the base $\overline{HT}$?

d. What is the equation of the perpendicular bisector of $\overline{HT}$?

e. How is the point A related to the perpendicular bisector of $\overline{HT}$?

2. A circle is centered at the origin and has radius r units. $\triangle PQR$ is a triangle such all three of its vertices belong to the circle and one of its sides is a diameter of the circle. We can place the x-axis along the diameter, as shown, to simplify the analysis. In this problem, $\overline{PQ}$ is a diameter of the circle with endpoints on the x-axis, and $R(b, c)$ is any other point on the circle as shown.

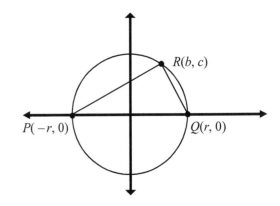

a. Write the equation for the circle.

b. Explain why $b^2 + c^2 = r^2$.

c. Find the slope of $\overline{PR}$.

d. Find the slope of $\overline{RQ}$.

e. Show that $\overline{PR}$ and $\overline{RQ}$ are perpendicular.

f. Explain why $\triangle PQR$ is a right triangle.

g. What can you say about any triangle such that all three of its vertices belong to a circle and one of the sides of the triangle is a diameter of the circle?

Activity 13.3.1 Translations, Reflections, and Rotations

Material: pen, ruler, protractor, and compass   Name _____

A **mapping** relates two sets of points in the plane. In a mapping, the **image** of point P is the point P', and the **preimage** of P' is P. A **transformation** maps the plane to itself such that every point in the plane has one image, and every point in the plane has one preimage. A **rigid motion** is a transformation that maps a line segment to a congruent line segment. This activity focuses on three basic rigid motions of the plane: translation (slide), reflection (flip), and rotation (turn). Every possible rigid motion of the plane can be expressed as some sequence of these three basic rigid motions, so we will focus on these three types of movements.

1. A translation maps point A to point A' in the coordinate system as shown.

    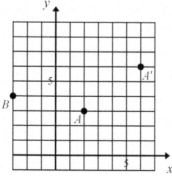

    a. Describe how points move under the translation.

    b. What are the coordinates of B'?

    c. Determine the slope of the line $\overleftrightarrow{AA'}$.

    d. Determine the slope of the line $\overleftrightarrow{BB'}$.

    e. Calculate and compare the distances AA' and BB'.

    f. Calculate and compare AB and $A'B'$.

2. A translation moves points 25 mm to the right and 15 mm down. Use a ruler to sketch the image of the triangle under this translation.

3. The diagram shows line l and points A and B. Use a pen to mark the points A and B. Fold the paper on the line l and press the paper so that the marked points create images. Label the images A' and B'. How are the lines l and $\overleftrightarrow{AA'}$ related?

    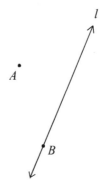

4. The diagram shows the reflection of A about the line of reflection l.

    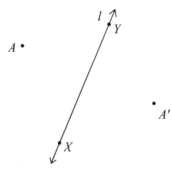

    a. Use a straightedge to construct the line segment $\overline{AA'}$ and let Q be the intersection of l and $\overline{AA'}$.

    b. use a ruler to measure the distances AQ and QA', then compare the distances.

    c. Measure $\angle AQX$.

5. Use Problems 3 and 4 to help complete the definition. A **reflection** is a transformation of the plane such that there is a line *l*, called the mirror line, with the properties:

   a. If the point *A* belongs to *l*, then _____.

   b. If the point *A* does not belong to *l*, then *l* is the
   _____ of $\overline{AA'}$.

6. Use a pencil and ruler to sketch the image of △*ABC* under a reflection about line *l*.
   (Hint: If you know that *X* does not belong to *l*, then you know that *l* is the perpendicular bisector of the line segment with endpoints *X* and its image *X'*.)

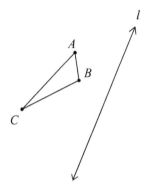

7. The diagram shows the image of $\overline{AB}$ under a reflection about a mirror line.

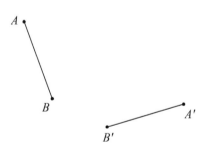

   a. Sketch the mirror line.

   b. Explain how you arrived at your answer.

8. The diagram shows point *A* and its image *A'* under a rotation with center *P* and counter-clockwise rotation of 30°.

   a. Set the radius of the compass to *PA* units. Use the compass to construct a circle with center *P* and radius *PA*.

   b. How are *PA* and *PA'* related?

   c. Measure ∠*APA'*.

   d. Use paper folding to create the perpendicular bisector of $\overline{AA'}$. What did you notice?

9. Use a compass, protractor, and straightedge to draw the image of $\overline{AB}$ under a clockwise rotation of 40°.

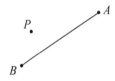

10. A rotation maps *A* to *A'* and *B* to *B'* as shown.

    a. Find the center of rotation.

    b. Determine the degree of rotation.

    $\overset{A}{\cdot}$ $\overset{\cdot}{B'}$

    $A'\!\cdot$ $\cdot B$

11. A clockwise rotation of 40° with center *P* maps a particular point to itself. What can you say about that point?

Activity 13.3.2   Size Transformations

Material: sharp pencil, ruler, and compass   Name _____

A **size transformation** (also called dilation) with center P and scale factor k (where $0 < k$) is a transformation such that
  a. The point P is mapped to P (the center is mapped to itself).
  b. If $Q \neq P$, then Q is mapped to the point Q' on $\overrightarrow{PQ}$ with $PQ' = k \cdot PQ$ (the image Q' belongs to the ray $\overrightarrow{PQ}$).

The following diagram shows how a size transformation with center P and scale factor 2 maps A to A' and B to B'.

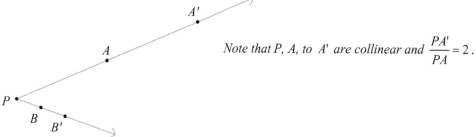

Note that P, A, to A' are collinear and $\dfrac{PA'}{PA} = 2$.

1. A size transformation has center P and scale factor $k = 2$, with A and B' as shown.
    a. Sketch A'.
    b. Sketch B.

    • B'
    • A
      • P

2. A size transformation has center P and scale factor $k = 1/2$, with A and B' as shown.

    a. Sketch A'.
    b. Sketch B.

    A •

    P •
          • B'

3. Determine the location of the center of the size transformation that maps $\overline{AB}$ to $\overline{A'B'}$.

    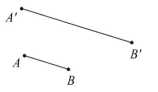

4. Do the following.
    a. Use a straightedge and compass to draw the image of A and B under a size transformation with center Q and scale factor 3.

    b. Measure and compare $A'B'$ and AB.

5. A size transformation maps $\overline{AB}$ to $\overline{A'B'}$.

    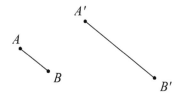

    a. Is the scale factor more than 1 or less than 1?
    b. Use a ruler to determine the scale factor.

6. A size transformation maps eyeglass A to eyeglass B. Use a ruler to determine the scale factor.

Activity 13.3.3   Creating Escher-like Tessellations

Material: sharp pencil, scissors, and tracing paper   Name _____

A **tessellation** is a repetition of a shape that covers the plane without any gaps or overlaps. The repeating shape is called a tile. A **tile** is a simple closed curve and its interior. Figure 1 illustrates that any triangle tessellates the plane.

Figure 1: Any triangular tessellates the plane.

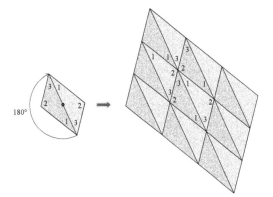

Figure 2 illustrates how to tessellate a plane using any convex quadrilateral (but the procedure also works for any concave quadrilateral). Pick a side of the quadrilateral. Then rotate the quadrilateral 180° about the midpoint of the side. Trace the image. Then pick a side of the image and rotate the image 180° about the midpoint of the side. Trace the new image. By repeating this process, the quadrilateral tessellates the plane.

Figure 2. Rotating the quadrilateral using a sequence of 180° rotations.

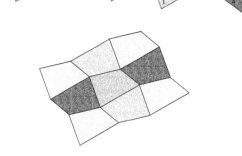

The Dutch artist Maurits Cornelis Escher (1898-1972) studied architecture and created pictures on paper and in woodcuts. He integrated geometry, illusion, and mathematics into his work, despite the fact he had no formal training in mathematics. He created tessellations with tiles in the shapes of people and animals. The official website for M.C. Escher is www.mcescher.com. An Escher-like tessellation is a tessellation based on interlocking figures.

Steps 1-6 illustrate how to make a tile for an Escher-like tessellation that uses two transformations to transform a rectangle. Use tracing paper to perform the translations on this worksheet (enjoy discovering how to use the tracing paper for the transformations). Step 7 shows a few tiles placed together without any gaps or overlaps.

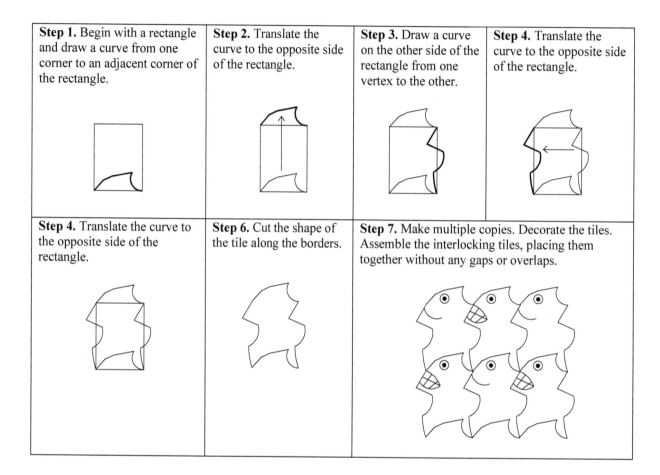

Step 1. Begin with a rectangle and draw a curve from one corner to an adjacent corner of the rectangle.	**Step 2.** Translate the curve to the opposite side of the rectangle.	**Step 3.** Draw a curve on the other side of the rectangle from one vertex to the other.	**Step 4.** Translate the curve to the opposite side of the rectangle.
Step 4. Translate the curve to the opposite side of the rectangle.	**Step 6.** Cut the shape of the tile along the borders.	**Step 7.** Make multiple copies. Decorate the tiles. Assemble the interlocking tiles, placing them together without any gaps or overlaps.	

1.   Follow these steps to make a tile for an Escher-like tessellation based on two translations. For more enjoyment, decorate the tile.

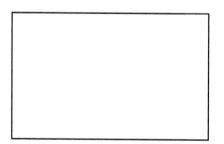

Steps 1-6 illustrate how to make a tile for an Escher-like tessellation that uses one reflection followed by two rotations to transform an equilateral triangle. Use tracing paper to perform the reflection and rotations on this worksheet (enjoy discovering how to use the tracing paper for the transformations). Step 7 shows a few tiles placed together without any gaps or overlaps.

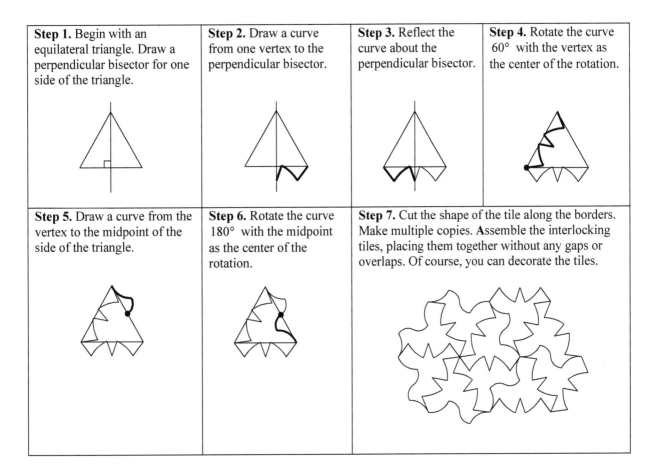

Step 1. Begin with an equilateral triangle. Draw a perpendicular bisector for one side of the triangle.	**Step 2.** Draw a curve from one vertex to the perpendicular bisector.	**Step 3.** Reflect the curve about the perpendicular bisector.	**Step 4.** Rotate the curve 60° with the vertex as the center of the rotation.
Step 5. Draw a curve from the vertex to the midpoint of the side of the triangle.	**Step 6.** Rotate the curve 180° with the midpoint as the center of the rotation.	**Step 7.** Cut the shape of the tile along the borders. Make multiple copies. Assemble the interlocking tiles, placing them together without any gaps or overlaps. Of course, you can decorate the tiles.	

2. Follow these steps to make a tile for an Escher-like tessellation based on one reflection followed by two rotations. For more enjoyment, decorate the tile.

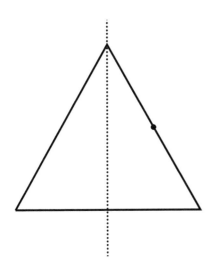

A1 **Base Ten Blocks**

A2 Two Color Chips

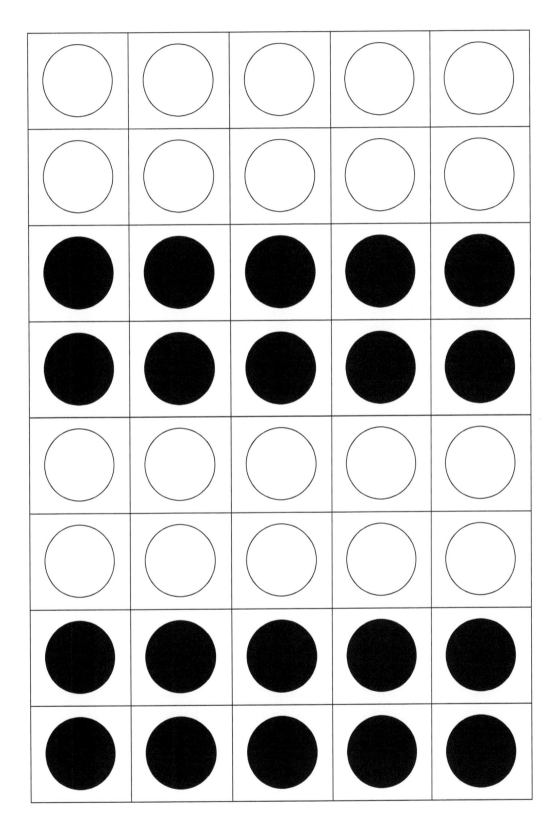

A3 **Pattern Blocks**

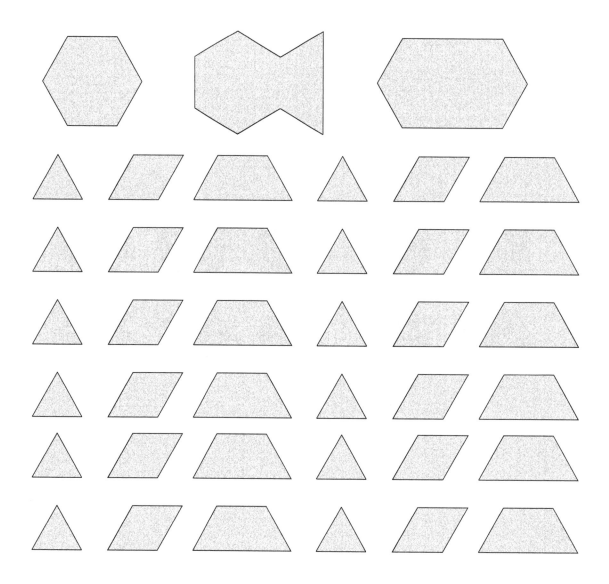

A4 **Fraction Strips**

$\frac{1}{2}$	$\frac{1}{2}$
$\frac{1}{2}$	$\frac{1}{2}$
$\frac{1}{2}$	$\frac{1}{2}$

$\frac{1}{3}$	$\frac{1}{3}$	$\frac{1}{3}$
$\frac{1}{3}$	$\frac{1}{3}$	$\frac{1}{3}$
$\frac{1}{3}$	$\frac{1}{3}$	$\frac{1}{3}$
$\frac{1}{3}$	$\frac{1}{3}$	$\frac{1}{3}$

$\frac{1}{4}$	$\frac{1}{4}$	$\frac{1}{4}$	$\frac{1}{4}$
$\frac{1}{4}$	$\frac{1}{4}$	$\frac{1}{4}$	$\frac{1}{4}$
$\frac{1}{4}$	$\frac{1}{4}$	$\frac{1}{4}$	$\frac{1}{4}$
$\frac{1}{4}$	$\frac{1}{4}$	$\frac{1}{4}$	$\frac{1}{4}$
$\frac{1}{4}$	$\frac{1}{4}$	$\frac{1}{4}$	$\frac{1}{4}$

$\frac{1}{5}$	$\frac{1}{5}$	$\frac{1}{5}$	$\frac{1}{5}$	$\frac{1}{5}$
$\frac{1}{5}$	$\frac{1}{5}$	$\frac{1}{5}$	$\frac{1}{5}$	$\frac{1}{5}$
$\frac{1}{5}$	$\frac{1}{5}$	$\frac{1}{5}$	$\frac{1}{5}$	$\frac{1}{5}$
$\frac{1}{5}$	$\frac{1}{5}$	$\frac{1}{5}$	$\frac{1}{5}$	$\frac{1}{5}$
$\frac{1}{5}$	$\frac{1}{5}$	$\frac{1}{5}$	$\frac{1}{5}$	$\frac{1}{5}$
$\frac{1}{5}$	$\frac{1}{5}$	$\frac{1}{5}$	$\frac{1}{5}$	$\frac{1}{5}$

198

A5 **Fraction Strips**

$\frac{1}{6}$	$\frac{1}{6}$	$\frac{1}{6}$	$\frac{1}{6}$	$\frac{1}{6}$	$\frac{1}{6}$
$\frac{1}{6}$	$\frac{1}{6}$	$\frac{1}{6}$	$\frac{1}{6}$	$\frac{1}{6}$	$\frac{1}{6}$
$\frac{1}{6}$	$\frac{1}{6}$	$\frac{1}{6}$	$\frac{1}{6}$	$\frac{1}{6}$	$\frac{1}{6}$
$\frac{1}{6}$	$\frac{1}{6}$	$\frac{1}{6}$	$\frac{1}{6}$	$\frac{1}{6}$	$\frac{1}{6}$
$\frac{1}{6}$	$\frac{1}{6}$	$\frac{1}{6}$	$\frac{1}{6}$	$\frac{1}{6}$	$\frac{1}{6}$
$\frac{1}{6}$	$\frac{1}{6}$	$\frac{1}{6}$	$\frac{1}{6}$	$\frac{1}{6}$	$\frac{1}{6}$
$\frac{1}{6}$	$\frac{1}{6}$	$\frac{1}{6}$	$\frac{1}{6}$	$\frac{1}{6}$	$\frac{1}{6}$

$\frac{1}{8}$	$\frac{1}{8}$	$\frac{1}{8}$	$\frac{1}{8}$	$\frac{1}{8}$	$\frac{1}{8}$	$\frac{1}{8}$	$\frac{1}{8}$
$\frac{1}{8}$	$\frac{1}{8}$	$\frac{1}{8}$	$\frac{1}{8}$	$\frac{1}{8}$	$\frac{1}{8}$	$\frac{1}{8}$	$\frac{1}{8}$
$\frac{1}{8}$	$\frac{1}{8}$	$\frac{1}{8}$	$\frac{1}{8}$	$\frac{1}{8}$	$\frac{1}{8}$	$\frac{1}{8}$	$\frac{1}{8}$
$\frac{1}{8}$	$\frac{1}{8}$	$\frac{1}{8}$	$\frac{1}{8}$	$\frac{1}{8}$	$\frac{1}{8}$	$\frac{1}{8}$	$\frac{1}{8}$
$\frac{1}{8}$	$\frac{1}{8}$	$\frac{1}{8}$	$\frac{1}{8}$	$\frac{1}{8}$	$\frac{1}{8}$	$\frac{1}{8}$	$\frac{1}{8}$
$\frac{1}{8}$	$\frac{1}{8}$	$\frac{1}{8}$	$\frac{1}{8}$	$\frac{1}{8}$	$\frac{1}{8}$	$\frac{1}{8}$	$\frac{1}{8}$
$\frac{1}{8}$	$\frac{1}{8}$	$\frac{1}{8}$	$\frac{1}{8}$	$\frac{1}{8}$	$\frac{1}{8}$	$\frac{1}{8}$	$\frac{1}{8}$
$\frac{1}{8}$	$\frac{1}{8}$	$\frac{1}{8}$	$\frac{1}{8}$	$\frac{1}{8}$	$\frac{1}{8}$	$\frac{1}{8}$	$\frac{1}{8}$
$\frac{1}{8}$	$\frac{1}{8}$	$\frac{1}{8}$	$\frac{1}{8}$	$\frac{1}{8}$	$\frac{1}{8}$	$\frac{1}{8}$	$\frac{1}{8}$

A6 **Fraction Strips**

$\frac{1}{10}$	$\frac{1}{10}$	$\frac{1}{10}$	$\frac{1}{10}$	$\frac{1}{10}$	$\frac{1}{10}$	$\frac{1}{10}$	$\frac{1}{10}$	$\frac{1}{10}$	$\frac{1}{10}$
$\frac{1}{10}$	$\frac{1}{10}$	$\frac{1}{10}$	$\frac{1}{10}$	$\frac{1}{10}$	$\frac{1}{10}$	$\frac{1}{10}$	$\frac{1}{10}$	$\frac{1}{10}$	$\frac{1}{10}$
$\frac{1}{10}$	$\frac{1}{10}$	$\frac{1}{10}$	$\frac{1}{10}$	$\frac{1}{10}$	$\frac{1}{10}$	$\frac{1}{10}$	$\frac{1}{10}$	$\frac{1}{10}$	$\frac{1}{10}$
$\frac{1}{10}$	$\frac{1}{10}$	$\frac{1}{10}$	$\frac{1}{10}$	$\frac{1}{10}$	$\frac{1}{10}$	$\frac{1}{10}$	$\frac{1}{10}$	$\frac{1}{10}$	$\frac{1}{10}$
$\frac{1}{10}$	$\frac{1}{10}$	$\frac{1}{10}$	$\frac{1}{10}$	$\frac{1}{10}$	$\frac{1}{10}$	$\frac{1}{10}$	$\frac{1}{10}$	$\frac{1}{10}$	$\frac{1}{10}$
$\frac{1}{10}$	$\frac{1}{10}$	$\frac{1}{10}$	$\frac{1}{10}$	$\frac{1}{10}$	$\frac{1}{10}$	$\frac{1}{10}$	$\frac{1}{10}$	$\frac{1}{10}$	$\frac{1}{10}$
$\frac{1}{10}$	$\frac{1}{10}$	$\frac{1}{10}$	$\frac{1}{10}$	$\frac{1}{10}$	$\frac{1}{10}$	$\frac{1}{10}$	$\frac{1}{10}$	$\frac{1}{10}$	$\frac{1}{10}$
$\frac{1}{10}$	$\frac{1}{10}$	$\frac{1}{10}$	$\frac{1}{10}$	$\frac{1}{10}$	$\frac{1}{10}$	$\frac{1}{10}$	$\frac{1}{10}$	$\frac{1}{10}$	$\frac{1}{10}$
$\frac{1}{10}$	$\frac{1}{10}$	$\frac{1}{10}$	$\frac{1}{10}$	$\frac{1}{10}$	$\frac{1}{10}$	$\frac{1}{10}$	$\frac{1}{10}$	$\frac{1}{10}$	$\frac{1}{10}$
$\frac{1}{10}$	$\frac{1}{10}$	$\frac{1}{10}$	$\frac{1}{10}$	$\frac{1}{10}$	$\frac{1}{10}$	$\frac{1}{10}$	$\frac{1}{10}$	$\frac{1}{10}$	$\frac{1}{10}$
$\frac{1}{10}$	$\frac{1}{10}$	$\frac{1}{10}$	$\frac{1}{10}$	$\frac{1}{10}$	$\frac{1}{10}$	$\frac{1}{10}$	$\frac{1}{10}$	$\frac{1}{10}$	$\frac{1}{10}$

A7 **Fraction Strips**

$\frac{1}{12}$	$\frac{1}{12}$	$\frac{1}{12}$	$\frac{1}{12}$	$\frac{1}{12}$	$\frac{1}{12}$	$\frac{1}{12}$	$\frac{1}{12}$	$\frac{1}{12}$	$\frac{1}{12}$	$\frac{1}{12}$	$\frac{1}{12}$
$\frac{1}{12}$	$\frac{1}{12}$	$\frac{1}{12}$	$\frac{1}{12}$	$\frac{1}{12}$	$\frac{1}{12}$	$\frac{1}{12}$	$\frac{1}{12}$	$\frac{1}{12}$	$\frac{1}{12}$	$\frac{1}{12}$	$\frac{1}{12}$
$\frac{1}{12}$	$\frac{1}{12}$	$\frac{1}{12}$	$\frac{1}{12}$	$\frac{1}{12}$	$\frac{1}{12}$	$\frac{1}{12}$	$\frac{1}{12}$	$\frac{1}{12}$	$\frac{1}{12}$	$\frac{1}{12}$	$\frac{1}{12}$
$\frac{1}{12}$	$\frac{1}{12}$	$\frac{1}{12}$	$\frac{1}{12}$	$\frac{1}{12}$	$\frac{1}{12}$	$\frac{1}{12}$	$\frac{1}{12}$	$\frac{1}{12}$	$\frac{1}{12}$	$\frac{1}{12}$	$\frac{1}{12}$
$\frac{1}{12}$	$\frac{1}{12}$	$\frac{1}{12}$	$\frac{1}{12}$	$\frac{1}{12}$	$\frac{1}{12}$	$\frac{1}{12}$	$\frac{1}{12}$	$\frac{1}{12}$	$\frac{1}{12}$	$\frac{1}{12}$	$\frac{1}{12}$
$\frac{1}{12}$	$\frac{1}{12}$	$\frac{1}{12}$	$\frac{1}{12}$	$\frac{1}{12}$	$\frac{1}{12}$	$\frac{1}{12}$	$\frac{1}{12}$	$\frac{1}{12}$	$\frac{1}{12}$	$\frac{1}{12}$	$\frac{1}{12}$
$\frac{1}{12}$	$\frac{1}{12}$	$\frac{1}{12}$	$\frac{1}{12}$	$\frac{1}{12}$	$\frac{1}{12}$	$\frac{1}{12}$	$\frac{1}{12}$	$\frac{1}{12}$	$\frac{1}{12}$	$\frac{1}{12}$	$\frac{1}{12}$
$\frac{1}{12}$	$\frac{1}{12}$	$\frac{1}{12}$	$\frac{1}{12}$	$\frac{1}{12}$	$\frac{1}{12}$	$\frac{1}{12}$	$\frac{1}{12}$	$\frac{1}{12}$	$\frac{1}{12}$	$\frac{1}{12}$	$\frac{1}{12}$
$\frac{1}{12}$	$\frac{1}{12}$	$\frac{1}{12}$	$\frac{1}{12}$	$\frac{1}{12}$	$\frac{1}{12}$	$\frac{1}{12}$	$\frac{1}{12}$	$\frac{1}{12}$	$\frac{1}{12}$	$\frac{1}{12}$	$\frac{1}{12}$
$\frac{1}{12}$	$\frac{1}{12}$	$\frac{1}{12}$	$\frac{1}{12}$	$\frac{1}{12}$	$\frac{1}{12}$	$\frac{1}{12}$	$\frac{1}{12}$	$\frac{1}{12}$	$\frac{1}{12}$	$\frac{1}{12}$	$\frac{1}{12}$
$\frac{1}{12}$	$\frac{1}{12}$	$\frac{1}{12}$	$\frac{1}{12}$	$\frac{1}{12}$	$\frac{1}{12}$	$\frac{1}{12}$	$\frac{1}{12}$	$\frac{1}{12}$	$\frac{1}{12}$	$\frac{1}{12}$	$\frac{1}{12}$
$\frac{1}{12}$	$\frac{1}{12}$	$\frac{1}{12}$	$\frac{1}{12}$	$\frac{1}{12}$	$\frac{1}{12}$	$\frac{1}{12}$	$\frac{1}{12}$	$\frac{1}{12}$	$\frac{1}{12}$	$\frac{1}{12}$	$\frac{1}{12}$
$\frac{1}{12}$	$\frac{1}{12}$	$\frac{1}{12}$	$\frac{1}{12}$	$\frac{1}{12}$	$\frac{1}{12}$	$\frac{1}{12}$	$\frac{1}{12}$	$\frac{1}{12}$	$\frac{1}{12}$	$\frac{1}{12}$	$\frac{1}{12}$

A8 **Algebra Tiles**

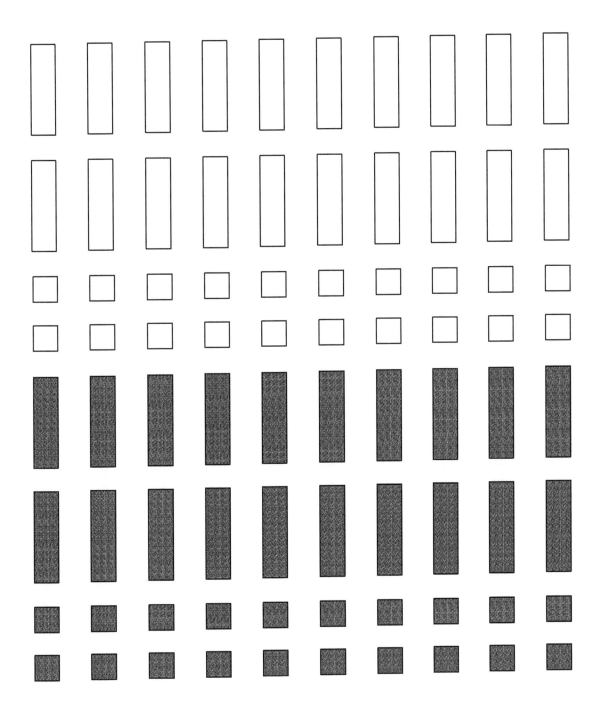

A9 **Graph Paper**

A10 Graph Paper

A11 **Graph Paper**

A12 **Table of Random Digits**

	1	2	3	4	5	6	7	8	9	10
1	10480	15011	01536	02011	81647	91646	69179	14194	62590	36207
	22368	46573	25595	85393	30995	89198	27982	53402	93965	34095
	24130	48360	22527	97265	76393	64809	15179	24830	49340	32081
	42167	93093	06243	61680	07856	16376	39440	53537	71341	57004
5	37570	39975	81837	16656	06121	91782	60468	81305	49684	60672
	77921	06907	11008	42751	27756	53498	18602	70659	90655	15053
	99562	72905	56420	69994	98872	31016	71194	18738	44013	48840
	96301	91977	05463	07972	18876	20922	94595	56869	69014	60045
	89579	14342	63661	10281	17453	18103	57740	84378	25331	12566
10	85475	36857	53342	53988	53060	59533	38867	62300	08158	17983
	28918	69578	88231	33276	70997	79936	56865	05859	90106	31595
	63553	40961	48235	03427	49626	69445	18663	72695	52180	20847
	09429	93969	52636	92737	88974	33488	36320	17617	30015	08272
	10365	61129	87529	85689	48237	52267	67689	93394	01511	26358
15	07119	97336	71048	08178	77233	13916	47564	81056	97735	85977
	51085	12765	51821	51259	77452	16308	60756	92144	49442	53900
	02368	21382	52404	60268	89368	19885	55322	44819	01188	65255
	01011	54092	33362	94904	31273	04146	18594	29852	71585	85030
	52162	53916	46369	58586	23216	14513	83149	98736	23495	64350
20	07056	97628	33787	09998	42698	06691	76988	13602	51851	46104
	48663	91245	85828	14346	09172	30168	90229	04734	59193	22178
	54164	58492	22421	74103	47070	25306	76468	26384	58151	06646
	32639	32363	05597	24200	13363	38005	94342	28728	35806	06912
	29334	27001	87637	87308	58731	00256	45834	15398	46557	41135

A13 **Net for Regular Pentagonal Prism**

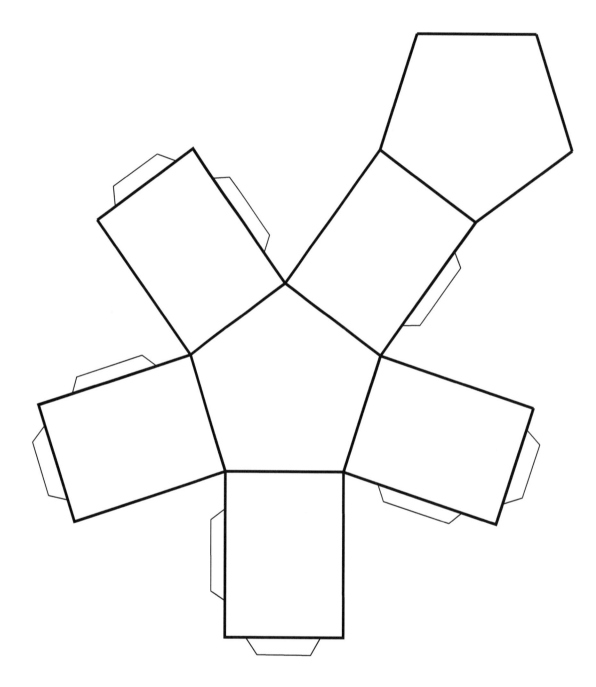

A14 **Net for Regular Pentagonal Pyramid**

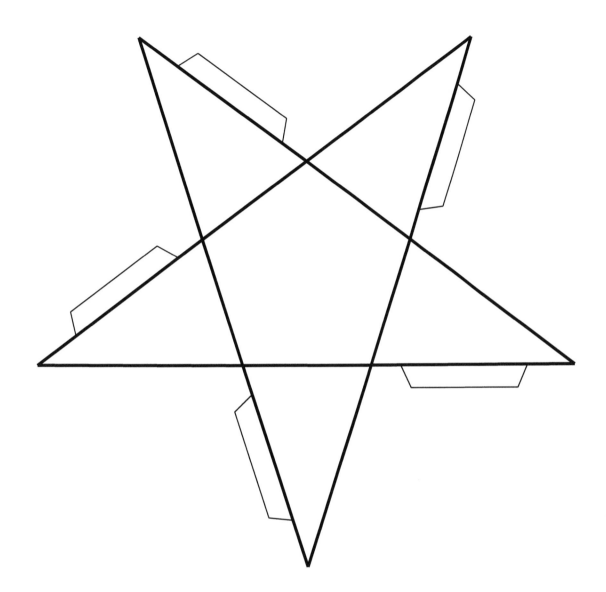

A15 **Circle**

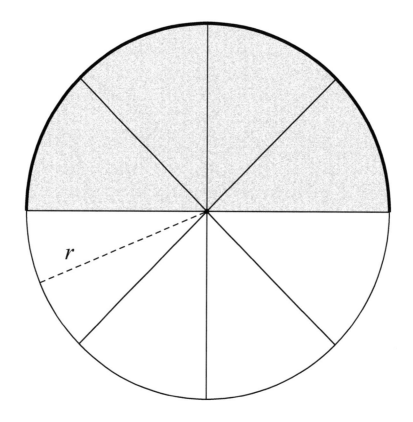

A16 **Line Segment Arrays**

A17 **Net for Pyramid**

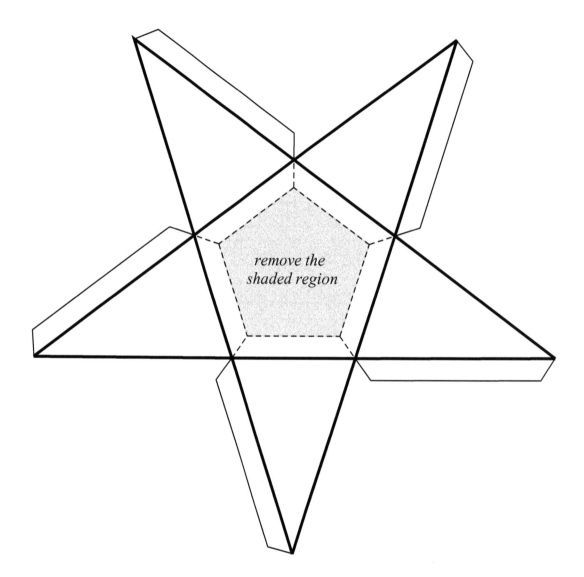

remove the shaded region

A18 **Net for Prism**

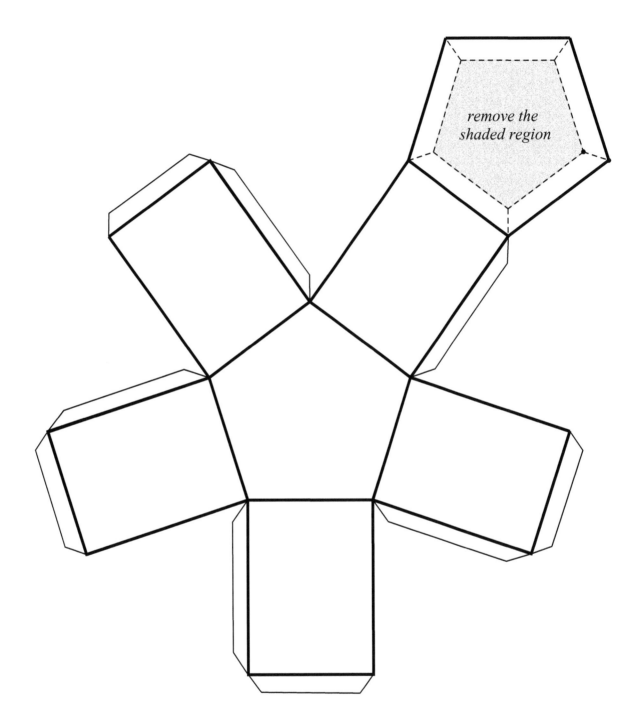

*remove the
shaded region*